CRITICAL THINKING & LOGICAL REASONING WORKBOOK-I

GIFT OF LOGIC™ SERIES

Boost Your Thinking Skills

An Essential Resource for Everyone

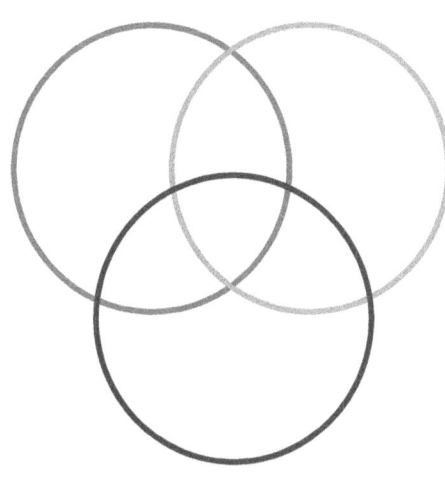

Verbal Reasoning
Analytical Reasoning
Pictorial Reasoning

THIRD EDITION

| FOR GRADES K-2 | STUDENTS, TEACHERS, AND PARENTS |

Ranga Raghuram

Gift Of Logic, Inc

http://www.giftoflogic.com
sales@giftoflogic.com

Critical Thinking and Logical Reasoning Workbook-1
ISBN-13: 978-1494830939
ISBN-10: 1494830930

Third Edition
1-2014

Copyright © 2009 Gift Of Logic, Inc. All rights reserved. No part of this publication may be reproduced, stored in a retrieval system, transmitted in any form or by any means, electronic, mechanical, photocopying, recording or otherwise, without the written permission of the publisher.

License: This book is licensed for use by one person only. Use of this book in a group setting (classroom, workshop, etc) without the written permission of the publisher is prohibited. Unauthorized duplication is strictly prohibited by law. Contact the publisher at sales@giftoflogic.com for classroom/school/group licensing.

GIFT OF LOGIC™
CRITICAL THINKING & LOGICAL REASONING CURRICULUM
12 WORKBOOKS TO BOOST YOUR THINKING SKILLS

For Kindergarten, Grade 1, and Grade 2

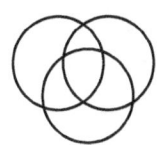
Workbook# 0

Verbal Reasoning	Finding the truth, Inferencing, Analogies, Synonyms and Antonyms, Agree/Disagree
Analytic Reasoning	Memory drill, Decision making, Positioning, Sudoku
Pictorial Reasoning	Connect the dots, Mazes, Picture Sequence, Spot the difference, etc

Workbook# 1

Verbal Reasoning	Finding the truth, Inferencing, Analogies, Synonyms and Antonyms, Agree/Disagree
Analytic Reasoning	Sorting, Positioning, Picking, Assorted problems, Numeric and Alphabetic Sudoku
Pictorial Reasoning	Picture Sequence, Spot the difference, Odd picture

Workbook# 2

Verbal Reasoning	Finding the truth, Classification, Direct and Inverse relationship, Inferencing, Analogies, Agree/Disagree
Analytic Reasoning	Sequencing, Scheduling, Strategy, Picking, etc
Pictorial Reasoning	Picture Analogy, Odd picture, Pattern matching, etc

For Grade 3, Grade 4, and Grade 5

Workbook# 3

Verbal Reasoning	Not, And, Or, If .. then, Conditional inferencing, Unconditional inferencing, Symbolic Logic
Analytic Reasoning	Lists, Sequencing, Grouping, Venn Diagrams, Graph logic, Number logic, Letter logic, Sudoku
Pictorial Reasoning	Picture sequence, Picture analogy, Odd picture, Picture difference, Pattern matching

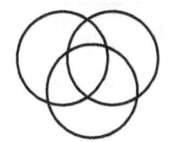
Workbook# 4

Verbal Reasoning	Contradiction, Converse, Inverse, Contrapositive, Conditional inferencing, Symbolic Logic
Analytic Reasoning	Scheduling, Looping, FIFO, LIFO, Correlation, Venn Diagram, Graph logic, Number logic, Sudoku, etc
Pictorial Reasoning	Picture sequence, Picture analogy, Odd picture, Picture difference, Pattern matching

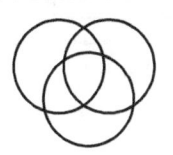
Workbook# 5

Verbal Reasoning	Biconditional, Categorical inferencing, Cause and Effect, Symbolic Logic, Agree/Disagree, Word and Sentence analogy
Analytic Reasoning	Correlation, Grouping, Venn Diagrams, Graph logic, Number logic, Letter logic, Sudoku, etc
Pictorial Reasoning	Picture sequence, Picture analogy, Odd picture, Picture difference, Pattern matching

********* Essential resource for everyone *********
*http://www.giftoflogic.com   *sales@giftoflogic.com

GIFT OF LOGIC™
CRITICAL THINKING & LOGICAL REASONING CURRICULUM
12 WORKBOOKS TO BOOST YOUR THINKING SKILLS

For Grades 6-12, College/University Students, Adults

Primer 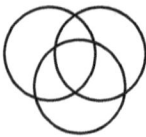 **Prereq**

Verbal Reasoning	Logical Operators, Conditional, Categorical and Causal reasoning, Validity, Fallacies, Symbolic Logic
Analytic Reasoning	Positioning, Grouping, Sudoku
Pictorial Reasoning	Pattern perception, Figure formation, Paper folding and cutting, Figure matrix, Rule detection

Workbook# 6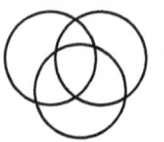

Verbal Reasoning	Arguments-Main point, Must be true, Cannot be true
Analytic Reasoning	Positioning, Grouping, Sudoku
Pictorial Reasoning	Pattern perception, Figure formation, Paper folding and cutting, Figure matrix, Rule detection

Workbook# 7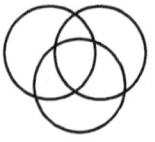

Verbal Reasoning	Arguments-Strengthening, Weakening
Analytic Reasoning	Positioning, Grouping, Sudoku
Pictorial Reasoning	Pattern perception, Figure formation, Paper folding and cutting, Figure matrix, Rule detection

Workbook# 8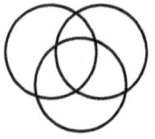

Verbal Reasoning	Arguments - Controversy, Paradox
Analytic Reasoning	Positioning, Grouping, Sudoku
Pictorial Reasoning	Pattern perception, Figure formation, Paper folding and cutting, Figure matrix, Rule detection

Workbook# 9

Verbal Reasoning	Arguments- Assumptions, Reasoning strategy
Analytic Reasoning	Positioning, Grouping, Sudoku
Pictorial Reasoning	Pattern perception, Figure formation, Paper folding and cutting, Figure matrix, Rule detection

Workbook# 10

Verbal Reasoning	Arguments-Flawed reasoning, Analogous reasoning
Analytic Reasoning	Positioning, Grouping, Sudoku
Pictorial Reasoning	Pattern perception, Figure formation, Paper folding and cutting, Figure matrix, Rule detection

********* Essential resource for everyone *********
Get the GIFT OF LOGIC™ today !
*http://www.giftoflogic.com   *sales@giftoflogic.com

Dear Reader:

Your decision to purchase this book is commendable. You now have in your hands, a comprehensive, easy-to-read book in Critical thinking and Logical reasoning that will introduce you to three different areas of thinking and reasoning - Verbal, Analytical and Pictorial. Solving problems in Verbal Reasoning is important to develop a critical mind. Solving problems in Analytic Reasoning is important to develop a flexible and resourceful mind. Solving problems in Pictorial Reasoning is important to develop a visually alert mind.

This book is presented in a workbook format to help you progress quickly. Parents and teachers are urged to complete the exercises ahead of the student and assist them whenever necessary with the help of detailed answers provided at the end of the book. This book can be used as a supplementary resource in the regular class room or it can be used during winter and summer vacations. College/University students, working professionals and retired individuals will also find the Gift Of Logic(tm) Series very useful in enhancing their problem solving abilities, confidence and general intellect.

Critical thinking and Logical reasoning must be practiced consistently to develop strong cognitive skills. After completing the exercises in this book, continue to read the other books in this series to get familiar with different types of Logical reasoning problems.

This workbook is one in a series of twelve workbooks. Please refer to the brochure before this page for a brief description of each workbook. Visit the website http://www.giftoflogic.com for more information.

Happy thinking and reasoning!

TABLE OF CONTENTS

Verbal Reasoning

Finding the truth..9

Word Analogy 12

Synonyms/ Antonyms 14

Agree or Disagree 16

Inferencing (Deductive Reasoning)
    Must be true..20
    Cannot be true...26

Analytic Reasoning

Sorting problems..29

Assorted problems..30

Positioning problems..38

Picking problems..43

TABLE OF CONTENTS

Analytic Reasoning (continued)

Sudoku
    Numeric Sudoku..48
    Alphabetic Sudoku...57

Pictorial Reasoning

Picture Sequence..67

Odd Picture..70

Spot the difference..73

Answers

Verbal..77

Analytic...88

Pictorial...123

Certificate of Completion

Name _____   Date _____

VERBAL REASONING

Name _____ Date _____

FINDING THE TRUTH

Finding the truth of statements is important for reasoning correctly.
Find the truth of the following statements and circle the correct answer.
Do some research if necessary.

1	All flowers are red in color. A) True B) False
2	We breathe in carbon dioxide and breathe out oxygen. A) True B) False
3	Sun is at the center of the Solar system. A) True B) False
4	Normal body temperature is 89.6 degrees Fahrenheit. A) True B) False
5	Doctors who take care of children only are called pediatricians. A) True B) False

Verbal Reasoning   Answers-77

Name _____   Date _____

FINDING THE TRUTH

Finding the truth of statements is important for reasoning correctly. Find the truth of the following statements and circle the correct answer. Do some research if necessary.

6	Earth has more land than water. A) True B) False
7	Alexander Graham Bell invented the telephone. A) True B) False
8	Plumbers fix problems with electricity. A) True B) False
9	When it rains, the roads are dry. A) True B) False
10	Ulan Bator is the capital of Mongolia. A) True B) False
11	An Ostrich is a bird that can run, but cannot fly. A) True B) False

Verbal Reasoning   Answers-77

© Gift Of Logic, Inc * Copying prohibited

FINDING THE TRUTH

Finding the truth of statements is important for reasoning correctly.
Find the truth of the following statements and circle the correct answer.
Do some research if necessary.

12	Sun is not at the center of the solar system. A) True B) False
13	Paris is not the capital of U.S.A. A) True B) False
14	Fish cannot survive on land. A) True B) False
15	Fill in the blanks below using "is" or "is not". Light _____ needed in order to see. To sing a song, it _____ necessary to be a famous singer.

Name _____ Date _____

WORD ANALOGY

The first two words separated by a colon (:) have a specific relationship. The third word has the same relationship with one of the words in the answer choices given. Circle the answer choice that will complete the analogy.

1	zoology : animals => botany : A) bottles B) plants
2	motorbike : ride => car : A) drive B) swim
3	book : read => guitar : A) sing B) play
4	rain : flood => sun : A) cool B) drought
5	affluent : rich => impoverished : A) improved B) poor
6	foreign : outside => native : A) above B) inside

Verbal Reasoning   Answers-78

© Gift Of Logic, Inc * Copying prohibited

Name _____   Date _____

WORD ANALOGY

The first two words separated by a colon (:) have a specific relationship. The third word has the same relationship with one of the words in the answer choices given. Circle the answer choice that will complete the analogy.

7	Singapore : Asia => Canada : A) North America B) Europe
8	English : Language => Painting : A) Art B) Song
9	Restaurant : Food => Theater : A) Movie B) Prayer
10	Stomach : Digest => Lung : A) Breathe B) Discharge
11	Light : See => Sound : A) Taste B) Hear
12	Roof : Foundation => Head : A) Hand B) Feet

Verbal Reasoning   Answers-78

© Gift Of Logic, Inc * Copying prohibited

Name _____   Date _____

SYNONYMS/ANTONYMS

Synonyms are words with the same meaning. Antonyms are words with opposite meaning. The following statements are made using synonyms or antonyms. Read each statements and decide if the statement is true or false.

1	To mute the sound is the same as increasing its volume. A) True B) False
2	To be enthusiastic is the same as being bored. A) True B) False
3	To be a quack is the same as being respected. A) True B) False
4	To be gullible is to be naive. A) True B) False
5	To reject something is not the same as discarding it. A) True B) False
6	To dodge responsibility is not the same as avoiding responsibility. A) True B) False

Verbal Reasoning   Answers-79
© Gift Of Logic, Inc * Copying prohibited

Name _____   Date _____

SYNONYMS/ANTONYMS

Synonyms are words with the same meaning. Antonyms are words with opposite meaning. The following statements are made using synonyms or antonyms. Read each statements and decide if the statement is true or false.

7	To invade is to trespass. A) True B) False
8	To fluctuate is to remain steady. A) True B) False
9	To retrieve something is to lose it. A) True B) False
10	To be feeble is not the same as being strong. A) True B) False
11	To fulfill a task is to finish it completely. A) True B) False
12	To shove something is the same as pulling it. A) True B) False
13	To snap something is the same as stretching it. A) True B) False

Verbal Reasoning   Answers-79
© Gift Of Logic, Inc * Copying prohibited

Name _____   Date _____

AGREE-DISAGREE

Read the statements of two people and find out whether they both agree or disagree with each other.

1

Dave: The big circle is outside the small circle.
Diane: The small circle is outside the big circle.

Dave and Diane
    A) agree with each other.
    B) disagree with each other.

2

Fred: This gold chain is expensive.
Harry: This gold chain is not cheap.

Fred and Harry
    A) have the same opinion about the gold chain.
    B) have different opinions about the gold chain.

3

Harry: This book is full of suspense. I was nervous while reading it.
Larry: I was not tense while reading this book, as it had only a little suspense.

Harry and Larry:
    A) agree about the amount of suspense in the book.
    B) disagree about the amount of suspense in the book.

AGREE-DISAGREE

4

Mark: Spending too much money is not a good habit.
Mary: Saving very little money is not a good habit.

Mark and Mary
  A) agree with each other.
  B) disagree with each other.

5

Rohan: This car is not fit for driving.
Randy: This car has safety problems.

Rohan and Randy
  A) agree with each other.
  B) disagree with each other.

6

Paul: Tall people live longer than short people.
Kevin: Being tall or short has nothing to do with how long people live.

Paul and Kevin
  A) agree with each other.
  B) disagree with each other.

AGREE-DISAGREE

7 some/all

Anita: Some insects bite.
Andrea: All insects bite.

Anita and Andrea
    A) agree with each other.
    B) disagree with each other.

8 many/few

Brad: Many cities in the world are very crowded.
Jennifer: Only a few cities in the world are very crowded.

Brad and Jennifer
    A) agree with each other.
    B) disagree with each other.

9 always/never

Neil: We should always explore the universe.
Lance: We should never stop exploring the universe.

Neil and Lance
    A) agree with each other.
    B) disagree with each other.

Name _____ Date _____

AGREE-DISAGREE

10   **at least/at most**

Cathy: At least five fire trucks are needed to stop the fire in this building.
Calvin: At most five trucks are needed to stop the fire in this building.

Cathy and Calvin
  A) agree with each other.
  B) disagree with each other.

---

11   **all/none**

Liz: All the students are in attendance today.
Eric: None of the students are absent today.

Liz and Eric
  A) agree with each other.
  B) disagree with each other.

---

12   **frequently/rarely**

Fred: I go to restaurants frequently.
Farida: You rarely go to restaurants.

Fred and Farida
  A) agree with each other.
  B) disagree with each other.

Name _____ Date _____

INFERENCE - must be true

Inferencing means finding from a given set of facts, information that is true, but not stated. In questions 1-12, some facts are given. Assuming that these facts are true, find out which of the given choices must be true.

---

1 cannot, can

Viral infections cannot be treated by antibiotics. Bacterial infections can be treated by antibiotics. Tracy has a throat infection caused by a bacteria called Streptococcus.

If the above statements are true, which one of the following must be true?
    A) Tracy's throat infection can be treated by antibiotics.
    B) Tracy's throat infection cannot be treated by antibiotics.

---

2 connect the facts

The policemen are looking for a man with a red shirt and blue pants. They want to arrest him for burglary. After searching for a long time, they finally arrested Mr. Rogers.

If the above information is true, then which one of the following facts must be true about Mr. Rogers?
    A) He was wearing a blue shirt and red pants.
    B) He was wearing a red shirt and blue pants.

Verbal Reasoning   Answers-84
© Gift Of Logic, Inc * Copying prohibited

Name —————————————— Date ——————————

| INFERENCE - must be true |

3
closer, warmer

The closer a planet is to the Sun, the warmer it is. Saturn is closer to Sun than Uranus.

Based on the information above, which one of the following must be true?
  A) Saturn is warmer than Uranus.
  B) Uranus is warmer than Saturn.

4

same

Javed is in the same class as Tina. Tina is in the same class as Randy.

Based on the information above, which one of the following must be true?
  A) Javed and Randy are in different classes.
  B) Javed and Randy are in the same class.

Name _____   Date _____

INFERENCE - must be true

5

or, and

For the Independence day, students are allowed to wear blue or green pants, and red or white shirts.

If the above information is true, which one of the following must be true?
    A) It is okay to wear a blue pant and a white shirt.
    B) It is okay to wear a green pant and a blue shirt.

6

one of

There are three lanes in the highway connecting NewCity and OldCity. But, one of the lanes is closed today due to construction.

If the above information is true, which one of the following must be true?
    A) The left most lane is open today.
    B) The left most lane is closed today.
    C) There are two lanes open today.

Name _____   Date _____

INFERENCE - must be true

7

except

Except Boris, everyone in his class submitted their homework. Everyone who submitted their homework got a prize.

From the above statements, we can infer that
      A) Boris got a prize.
      B) Everyone except Boris got a prize.

8

in addition

In addition to Javed, three others got gold medals.

From the above statement, we can infer that
      A) Javed is the only one to get a gold medal.
      B) Javed is not the only one to get a gold medal.

Name _____ Date _____

INFERENCE - must be true

9

logical chain (connect the facts)

People who perform tricks are smart. Magicians perform tricks.

If the above statements are true, which one of the following statements must be true?

    A) Magicians are not smart.
    B) Magicians are smart.

10

logical chain (connect the facts)

Boats with a small fuel tank cannot travel a long distance. Small boats have a small fuel tank.

If the above statements are true, which one of the following statements must be true?

    A) Small boats can travel long distance.
    B) Small boats can travel short distances only.

Name _____   Date _____

INFERENCE - must be true

11

logical chain (connect the facts)

Cats eat rats. Rats eat bugs.

If the above statements are true, which one of the following statements must be true?

    A) Cats eat bugs.
    B) Bugs eat rats.

12

logical chain (connect the facts)

We get sunlight from Sun. We get energy from sunlight.

If the above statements are true, which one of the following statements must be true?

    A) We get energy from Sun.
    B) We get sunlight from energy.

Name _____ Date_____

| INFERENCE - cannot be true |

You should know how to spot an incorrect inference. In questions 13-16, some facts are given. Assuming that these facts are true, find out which of the choices listed cannot be true. Use your common sense to answer these questions.

13

made with

Cake is made with eggs. The cost of eggs is going to increase soon.

If the above statements are true, which one of the following statements cannot be true?

    A) The cost of cake will decrease soon.
    B) The cost of cake will increase soon.

14

logical chain

Driving in the wrong lane is illegal. Illegal driving is dangerous.

If the above statements are true, which one of the following statements cannot be true?

    A) Driving in the wrong lane is not dangerous.
    B) Driving in the wrong lane is dangerous.

| INFERENCE - cannot be true |

15

everyone

Martin is four feet tall. Everyone taller than three feet can dive in the swimming pool.

If the above statements are true, which one of the following statements cannot be true?

    A) Martin cannot dive in the swimming pool.
    B) Martin can dive in the swimming pool.

16

mix

When water and ink are mixed, water takes the color of the ink. Red ink was mixed with water.

If the above statements are true, which one of the following statements cannot be true?

    A) Water turned blue.
    B) Water turned red.

ANALYTICAL REASONING

SORTING PROBLEMS

1
New York
New Delhi
Tokyo
London

The list shown above is sorted alphabetically in ascending order.
   A) True   B) False

2
Tracy
Victor
Wasim
Zachary

The list shown above is sorted alphabetically in ascending order.
   A) True   B) False

First Name	Last Name
Steve	Chaplin
Tom	Ewing
Abdul	Kazi
Nilesh	Wadia

3
The list shown is sorted alphabetically in ascending order by last name.
   A) True B) False

4

The list shown is sorted alphabetically in descending order by first name.
   A) True B) False

ASSORTED PROBLEMS — SEQUENCE

1

Cities named A, B, C, D, E, F, G, H, I, and J are connected as described below.

A to B, B to C and C to D.
B to E, E to F and F to G.
F to H, H to I.
I to J.
G to I.

Based on the information given above, reason and answer the following questions.

1) To go from A to H, we must go through G.
    A) True   B) False

2) To go from A to H, we must go through F.
    A) True   B) False

3) To go from A to I, we must go through H.
    A) True   B) False

4) The number of ways in which we can go from A to J is
    A) 1   B) 2

Analytical Reasoning   Answers-89

ASSORTED PROBLEMS   SEQUENCE

2

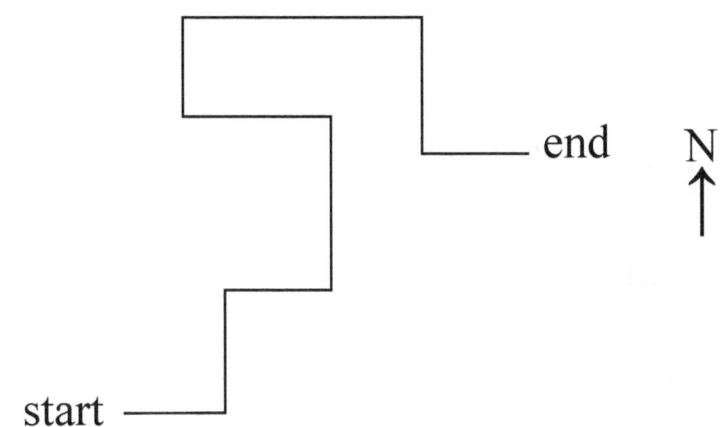

The figure above shows the roads to take from start to end. To go east or west, you must take the red bus. To go north or south, you must take the green bus.

1) How many green buses should you take to go from start to end?
   A) 3   B) 4

2) How many red buses should you take to go from start to end?
   A) 4   B) 5

3) How many buses should you take to go from start to end?
   A) 4   B) 5   C) 9

Name _____   Date _____

ASSORTED PROBLEMS — SWAP

3

Swapping means exchanging.

Birds	Animals
parrot	flamingo
tiger	dove
eagle	peacock
lion	fox

Swap the animals and birds in the groups shown above so that they are in the correct groups. After the swap, how many birds are there in the Birds group?

    A) 3   B) 5

4

A book shop carried the following advertisement in the local newspaper.

Swap 3 old books with 2 new books! Hurry! Deal ends soon!

1) Jane took 9 old books to the book shop. How many new books did she get?

    A) 9   B) 6

2) Jack took 5 old books to the book shop. How many new books did he get?

    A) 2   B) 3

Name _____ Date _____

ASSORTED PROBLEMS   SWAP

5

Foreign exchange rates are shown in the following table.

Foreign Exchange for 1 U.S. Dollar	
British Pounds	2
Indian Rupees	40
Swedish Krona	7

1) Meghan went to India to see the Taj Mahal. She went to the bank to exchange 10 U.S dollars. What did the bank swap with her?
   A) 40 Rupees   B) 400 Rupees

2) Mathew went to London to see his uncle. He gave the bank 100 U.S dollars. What did the bank exchange with him?
   A) 2 Pounds   B) 200 Kronas   C) 200 Pounds

6

Swap the wires in the left column so that their colors match their slots in the right column. Write the names of the wires in the blank column.

red wire		yellow slot
blue wire		green slot
green wire		blue slot
yellow wire		red slot

1) How many swaps does it take to put the wires in their correct slots?
   A) 4   B) 2

ASSORTED PROBLEMS — SWAP

7

position	object	swap
1	apple	
2	orange	
3	banana	

Swap the positions of apple and banana and fill the column titled "swap".

Swap Amber's position with Brendan's. Swap Drew's position with Kate's. Write the new positions in the blank table.

8

Amber	Drew
Kate	Brendan

9

8	7	9
t	4	s
a	f	m

Swap the row that has numbers only with the row that has alphabets only and write the new positions in the blank table below.

ASSORTED PROBLEMS   REVERSE

10

A book was bound incorrectly with the following page numbers from beginning to end: 6, 5, 4, 3, 2, 1

To fix the problem, the correct strategy would be
    A) to turn the pages upside down and bind again.
    B) to reverse the order of the pages and bind again.

11

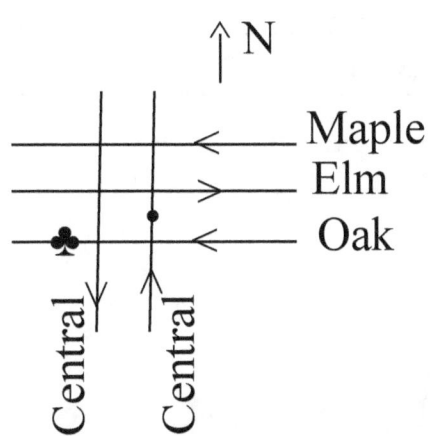

A network of one-way roads is shown in the figure above. Andrew is driving north on Central Avenue and is currently at the point indicated by • past Oak street. To reach the landmark indicated by the symbol ♣, Andrew must

    A) Reverse at Central and Elm.
    B) Reverse at Maple and Central.

Name _____   Date _____

| ASSORTED PROBLEMS | ORDER |

12

The following table shows the tasks need to be completed to build a house, but they are not in the logical sequence. Plan the correct sequence of tasks and write the correct order in the column titled Task#.

Task#	Tasks
	construct walls and roof
	lay foundation
	install windows and doors
	install carpets
	paint outside

13

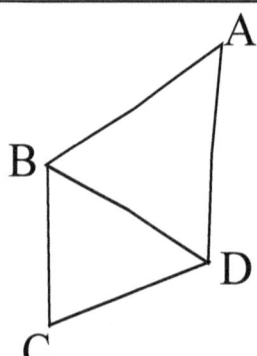

Cities A,B,C, and D are connected as shown. Chris lives in C and plans to visit his aunts Anne, Baker, and Drew who live in A, B, and D respectively. Aunt Anne wants him to carry a bag to give to aunt Drew. Aunt Drew wishes to add something to this bag and wants Chris to carry it to aunt Baker.

The correct order that Chris should follow is
    A) CD, DB, BA, AD, DC
    B) CB, BA, AD, DB, BC

ASSORTED PROBLEMS   SCHEDULE

14   schedule conflict

Zoya's schedule for the evening is to be decided from the events shown:
- TV : 5 PM - 5:30 PM
- Soccer : 6 PM - 7 PM
- Movie : 6:30 PM - 8:30 PM

1) Zoya does not want to skip soccer. What must she skip to schedule her evening without any conflicts?
     A) TV   B) Movie

2) If Zoya wants to see the movie, then what event must be skipped to keep her evening schedule without any conflicts?
     A) TV   B) Soccer

15   schedule conflict

Math: 8 AM - 9 AM; Reading: 9 AM - 10 AM; Logic 10 AM - 11 AM

Victor's schedule for Saturday is shown above.

1) His uncle wants to drop by at 10:20 AM, but can postpone his visit if Victor is busy. His uncle must postpone his visit by
     A) 30 minutes   B) 1 hour

2) Victor's friend wants to play for 45 minutes starting at 8 AM. His friend must advance his visit to
     A) 7:15 AM   B) 8:15 AM

Name _____   Date _____

POSITIONING PROBLEMS   FIXED POSITION

1

```
 ⊓  ⊓  ⊓
 1  2  3
```

Tom, Dick, and Harry must be seated in three stools. One in each stool. Tom must sit in stool #1.

Which of the following seatings are correct?

A) Dick   Tom   Harry
```
 ⊓    ⊓    ⊓
 1    2    3
```

B) Tom   Dick   Harry
```
 ⊓    ⊓    ⊓
 1    2    3
```

C) Tom   Harry   Dick
```
 ⊓    ⊓    ⊓
 1    2    3
```

Can Harry sit in stool# 1?
    A) Yes   B) No

Analytical Reasoning   Answers-95
© Gift Of Logic, Inc * Copying prohibited

Name _____ Date_____

POSITIONING PROBLEMS   OR

2

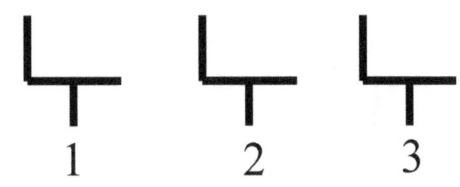

Andy, Brittany, and Cathy must be seated in the three chairs; only one in each chair. Andy can sit only in chair# 1 or chair# 3.

Which of the following seatings are correct?

A) Brittany Andy Cathy
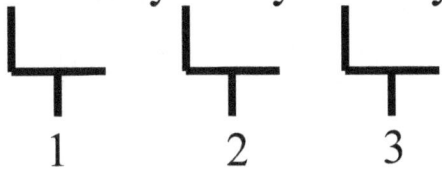

B) Andy Brittany Cathy
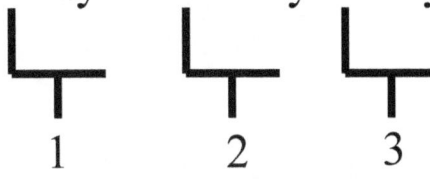

C) Cathy Brittany Andy
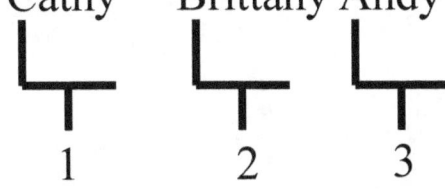

D) Andy Brittany Andy
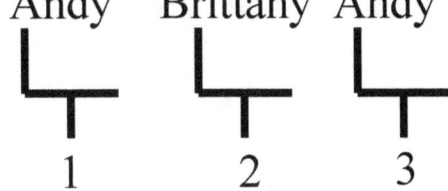

Andy can sit in any of the three chairs.
    A) True   B) False

Analytical Reasoning   Answers-96
© Gift Of Logic, Inc * Copying prohibited

POSITIONING PROBLEMS — LEFT OF

3

    1   2   3

Andy, Brittany, and Cathy must be seated in the three chairs. Andy must sit to the left of Brittany.

Which of the following seatings are correct?

A) Brittany (1)   Andy (2)   Cathy (3)

B) Andy (1)   Brittany (2)   Cathy (3)

C) Andy (1)   Cathy (2)   Brittany (3)

POSITIONING PROBLEMS   RIGHT OF

4
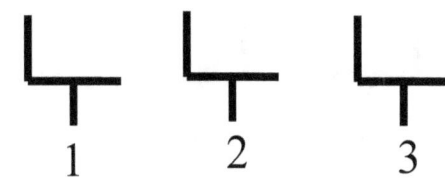

Paul, Queenie, and Robert must be seated in the three chairs; only one in each chair. Queenie must sit to the right of Robert.

Which of the following seatings are correct?

A) Paul   Queenie   Robert

B) Robert   Queenie   Paul
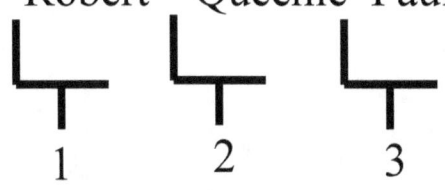

C) Robert   Paul   Queenie
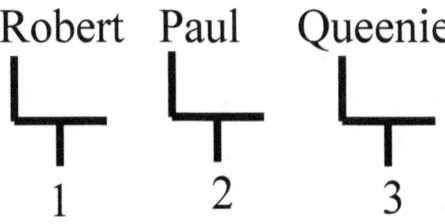

Can Queenie sit in chair # 1?
    A) Yes   B) No

POSITIONING PROBLEMS   BETWEEN

5

1   2   3

Larry, Margie, and Nina are to be seated in three chairs shown above. Larry must sit between Margie and Nina.

Which of the following seatings are correct?

A) Larry   Margie   Nina
     1   2   3

B) Nina   Larry   Margie
     1   2   3

C) Margie Larry   Nina
     1   2   3

Larry can sit only in chair # 2.
    A) True   B) False

Name _____   Date _____

PICKING PROBLEMS   MUST

1

The group below shows five students and the grades to which they belong.

Larry	Kindergarten
Mary	First grade
Nancy	Second grade
Oliver	First grade
Patrick	Kindergarten

From the above group of students, a team of three students is to be picked. Two of them must be in kindergarten.

Which of the following choices represents a correct team?

A) Larry, Mary, Oliver
B) Larry, Patrick, Oliver
C) Mary, Nancy, Oliver

Pick a team of three students from the above group. One student must be in the second grade.

Which of the following choices represents a correct team?

A) Nancy, Larry, Mary
B) Mary, Oliver, Patrick

Analytical Reasoning   Answers-100
© Gift Of Logic, Inc * Copying prohibited

PICKING PROBLEMS   MUST, EACH

2

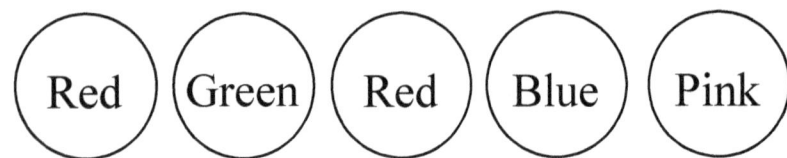

Five balls are available in the colors shown above.
A set of three balls must be picked from the five.
Each ball must be of different color.

Which of the following choices represent a correct set?

A)

B)

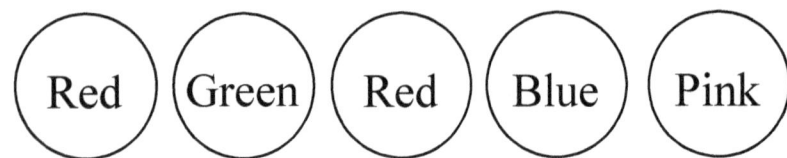

Five balls are available in the colors shown above.
A set of three balls must be picked. Two balls must be of the same color.
Which of the following choices represent a correct set?

A)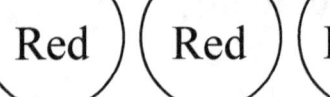

B)

Name ______________________ Date______________________

PICKING PROBLEMS TOGETHER

3

The group below shows four students:

Andy, Brittany, Cathy, Derek

Pick a team of three students. Andy and Brittany must be picked together.

Which of the following choices represents a correct team?

A) Andy, Brittany, Cathy

B) Andy, Cathy, Derek

C) Derek, Brittany

D) Cathy, Andy

E) Brittany, Andy, Derek

© Gift Of Logic, Inc * Copying prohibited

Name —————————————— Date ——————————

PICKING PROBLEMS   MUST

4

P, Q, R, S, and T are the names of five students.

Two teams, team A and team B are to be formed out of these five students.

Team A must have three students.

Team B must have two students.

P and S must be in Team A.

Q and R must be in team B.

Which of the following choices represent the correct teams?

	Team A	Team B
A)	P, S, T	Q, R
B)	P, S	Q, R, T
C)	Q, R, T	P, S

PICKING PROBLEMS   MUST, CANNOT

5

L, M, N, O, and P are the names of five students.

Two teams, team A and team B are to be formed out of these five students.

Team B must have two students.

M and N must be in the same team.

L and O cannot be in the same team.

Which of the following choices represent the correct teams?

	Team A	Team B
A)	M, N, O	L, P
B)	M, P, L	N, O
C)	L, N, O	M, P
D)	M, N	P

NUMERIC SUDOKU

Solve the Sudokus shown below. A solved Sudoku has numbers 1,2,3, and 4 appearing in each row, each column and the four bolded squares only once. You develop valuable positioning skills while solving these Sudokus.

1

1		4	3
4	3		1
2	1		4
3		1	2

2

4	3	1	
2		3	
1		2	3
	2		1

Name ______________________ Date______________________

NUMERIC SUDOKU

Solve the Sudokus shown below. A solved Sudoku has numbers 1,2,3, and 4 appearing in each row, each column and the four bolded squares only once. You develop valuable positioning skills while solving these Sudokus.

3

	3	4	2
	4	1	3
3		2	
4	2	3	

4

2	1	4	3
4		1	2
1		3	
3	4	2	

© Gift Of Logic, Inc * Copying prohibited

Name _____   Date _____

NUMERIC SUDOKU

Solve the Sudokus shown below. A solved Sudoku has numbers 1,2,3, and 4 appearing in each row, each column and the four bolded squares only once. You develop valuable positioning skills while solving these Sudokus.

5

2	1	4	3
	3		
3	2	1	4
	4	3	

6

3			4
1	4		2
4			1
2		4	3

NUMERIC SUDOKU

Solve the Sudokus shown below. A solved Sudoku has numbers 1,2,3, and 4 appearing in each row, each column and the four bolded squares only once. You develop valuable positioning skills while solving these Sudokus.

7

4	3	1	2
2			
3	2	4	1
1			3

8

2		4	1
1	4		3
	1		2
3	2		4

NUMERIC SUDOKU

Solve the Sudokus shown below. A solved Sudoku has numbers 1,2,3, and 4 appearing in each row, each column and the four bolded squares only once. You develop valuable positioning skills while solving these Sudokus.

9

			3
1	3	2	
		3	
3			1

10

			4
3			
	1	2	

NUMERIC SUDOKU

Solve the Sudokus shown below. A solved Sudoku has numbers 1,2,3, and 4 appearing in each row, each column and the four bolded squares only once. You develop valuable positioning skills while solving these Sudokus.

11

3		4	
1	4		
	1	2	3
	3	1	

12

	4		
	2	3	4
2	3	4	
		2	

Analytical Reasoning   Answers-110
© Gift Of Logic, Inc * Copying prohibited

Name _____ Date _____

NUMERIC SUDOKU

Solve the Sudokus shown below. A solved Sudoku has numbers 1,2,3, and 4 appearing in each row, each column and the four bolded squares only once. You develop valuable positioning skills while solving these Sudokus.

13

	2		
1			4
		1	
3			2

14

3	4		
2			3
1			4
		1	

Name _____   Date_____

NUMERIC SUDOKU

Solve the Sudokus shown below. A solved Sudoku has numbers 1,2,3, and 4 appearing in each row, each column and the four bolded squares only once. You develop valuable positioning skills while solving these Sudokus.

15

	3		
	4	2	3
4	2	3	
		4	

16

3			2
	4	1	
	2	3	
1			4

NUMERIC SUDOKU

Solve the Sudokus shown below. A solved Sudoku has numbers 1,2,3, and 4 appearing in each row, each column and the four bolded squares only once. You develop valuable positioning skills while solving these Sudokus.

17

	2		
	3	1	2
2	4	3	
		2	

18

4	2		3
			2
2			
1	3		4

ALPHABETIC SUDOKU

Solve the Sudokus shown below. A solved Sudoku has alphabets A,B,C, and D appearing in each row, each column and the four bolded squares only once. You develop valuable positioning skills while solving these Sudokus.

1

	B		
			A
B		C	
	D		

2

			B
B		C	
	D		
C			A

Analytical Reasoning   Answers-114

ALPHABETIC SUDOKU

Solve the Sudokus shown below. A solved Sudoku has alphabets A,B,C, and D appearing in each row, each column and the four bolded squares only once. You develop valuable positioning skills while solving these Sudokus.

3

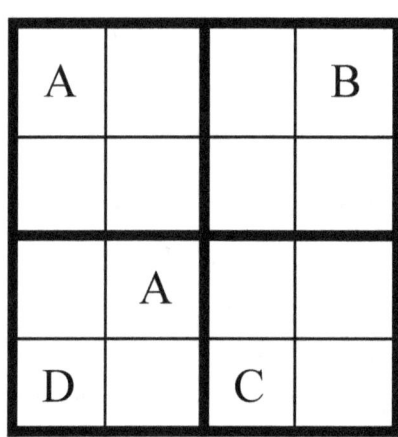

4

Analytical Reasoning   Answers-115

© Gift Of Logic, Inc * Copying prohibited

Name _____   Date _____

ALPHABETIC SUDOKU

Solve the Sudokus shown below. A solved Sudoku has alphabets A, B, C, and D appearing in each row, each column and the four bolded squares only once. You develop valuable positioning skills while solving these Sudokus.

5

B			C
	C		
	D	C	

6

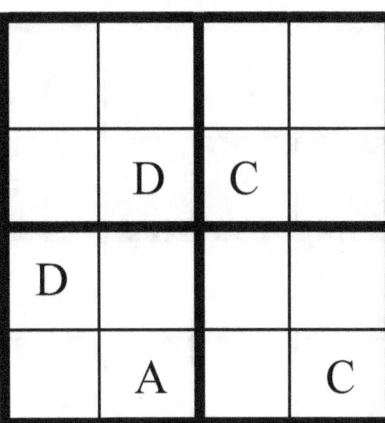

	D	C	
D			
	A		C

ALPHABETIC SUDOKU

Solve the Sudokus shown below. A solved Sudoku has alphabets A, B, C, and D appearing in each row, each column and the four bolded squares only once. You develop valuable positioning skills while solving these Sudokus.

7

D			
		C	
			A
A	D		

8

	C		A
A			
			B
C		A	

ALPHABETIC SUDOKU

Solve the Sudokus shown below. A solved Sudoku has alphabets A,B,C, and D appearing in each row, each column and the four bolded squares only once. You develop valuable positioning skills while solving these Sudokus.

9

			C
A		B	
		C	
C			A

10

			D
C			
	A	B	
		D	

Analytical Reasoning   Answers-118
© Gift Of Logic, Inc * Copying prohibited

Name _____   Date _____

ALPHABETIC SUDOKU

Solve the Sudokus shown below. A solved Sudoku has alphabets A,B,C, and D appearing in each row, each column and the four bolded squares only once. You develop valuable positioning skills while solving these Sudokus.

11

C		D	
A		C	
	A		C
	C		D

12

	D		
	B	C	D
B	C	D	
		B	

ALPHABETIC SUDOKU

Solve the Sudokus shown below. A solved Sudoku has alphabets A,B,C, and D appearing in each row, each column and the four bolded squares only once. You develop valuable positioning skills while solving these Sudokus.

13

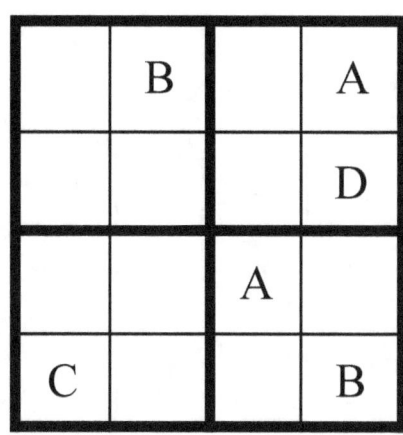

14

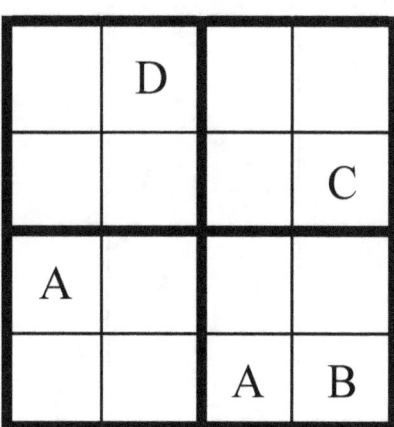

ALPHABETIC SUDOKU

Solve the Sudokus shown below. A solved Sudoku has alphabets A,B,C, and D appearing in each row, each column and the four bolded squares only once. You develop valuable positioning skills while solving these Sudokus.

15

	C		
	D	B	C
D	B	C	
		D	

16

C			B
	D	A	
	B	C	
A			D

Name ______________________ Date______________________

ALPHABETIC SUDOKU

Solve the Sudokus shown below. A solved Sudoku has alphabets A,B,C, and D appearing in each row, each column and the four bolded squares only once.You develop valuable positioning skills while solving these Sudokus.

17

	B		
	C	A	B
B	D	C	
		B	

18

D	B		C
			B
B			
A	C		D

© Gift Of Logic, Inc * Copying prohibited

Name _____   Date _____

PICTORIAL REASONING

Name —————————————— Date ——————————————

PICTURE SEQUENCE

Figure out the logic in the sequence of pictures shown, and draw the next picture in the sequence.

1

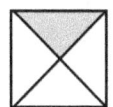 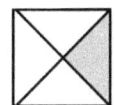   ?

2

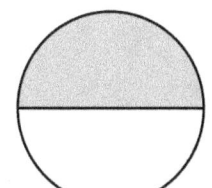 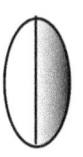 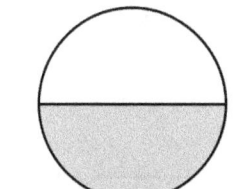   ?

3

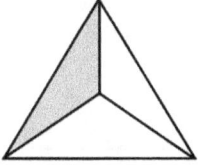 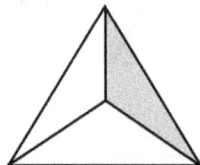   ?

4

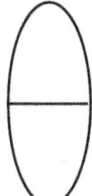 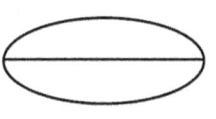   ?

Pictorial Reasoning   Answers-123   67
© Gift Of Logic, Inc * Copying prohibited

Name ———————————— Date ————————————

PICTURE SEQUENCE

Figure out the logic in the sequence of pictures shown, and draw the next picture in the sequence that will continue the logic.

5

6

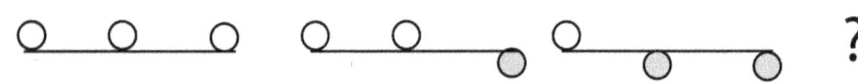

7

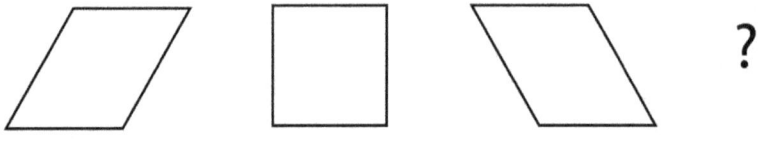

8

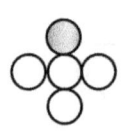 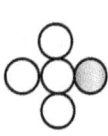 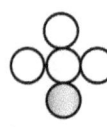 ?

Pictorial Reasoning   Answers-123

PICTURE SEQUENCE

Figure out the logic in the sequence of pictures shown, and draw the next picture in the sequence that will continue the logic.

9

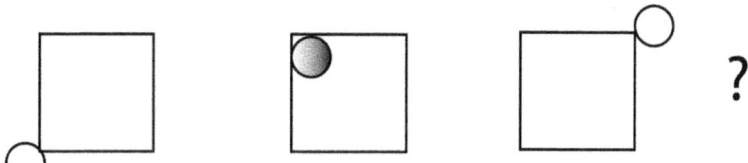

10

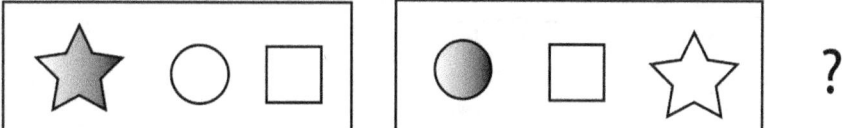

11

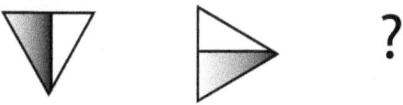

12

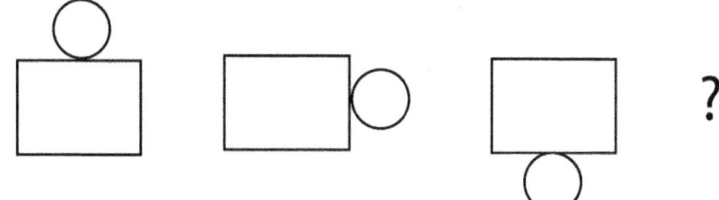

Name —————————————— Date ——————————

ODD PICTURE

In each question below, find the odd picture and circle the answer.

1

    A   B   C

2

    A   B   C

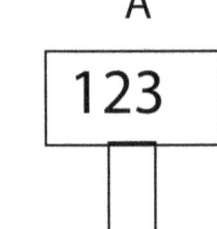 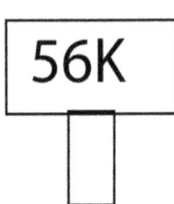 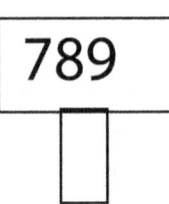

3

    A   B   C

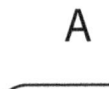 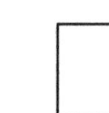 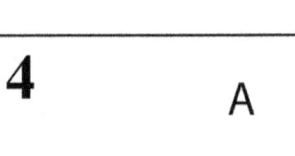

4

    A   B   C

Pictorial Reasoning   Answers-124

© Gift Of Logic, Inc * Copying prohibited

Name —————————————— Date ——————————

ODD PICTURE

In each question below, find the odd picture and circle the answer.

5

    A   B   C

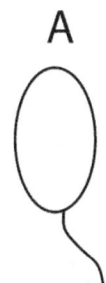  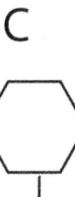

6

    A   B   C

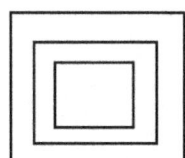  

7

    A   B   C

8

    A   B   C

 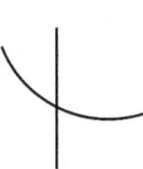

Pictorial Reasoning   Answers-124

© Gift Of Logic, Inc * Copying prohibited

Name _____ Date _____

ODD PICTURE

In each question below, find the odd picture and circle the answer.

9   A   B   C

 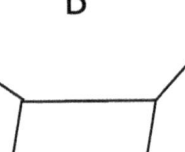 

10   A   B   C

11   A   B   C

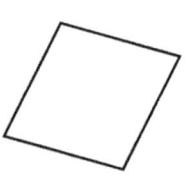 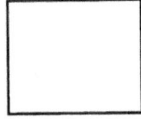 

12   A   B   C

  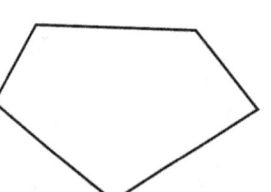 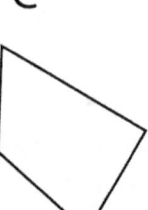

Pictorial Reasoning   Answers-124

© Gift Of Logic, Inc * Copying prohibited

Name _____   Date _____

SPOT THE DIFFERENCE

Which of the following figures are reflected, and which ones are scaled?

1

2

3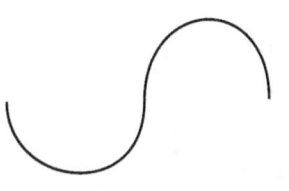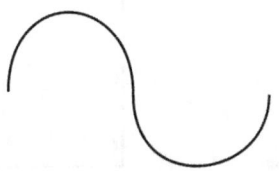

4

Pictorial Reasoning   Answers-125

Name _____   Date _____

SPOT THE DIFFERENCE

Spot the difference between the two figures in the questions below.

5

 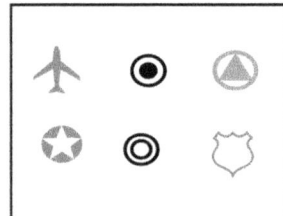

6

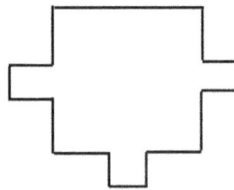 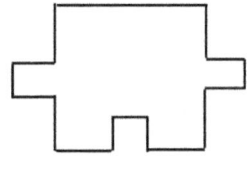

7

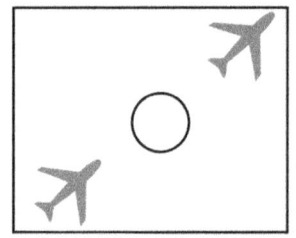 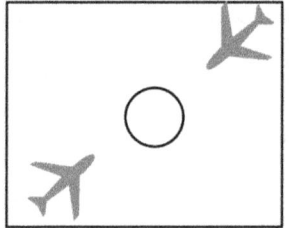

8

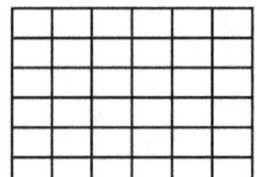 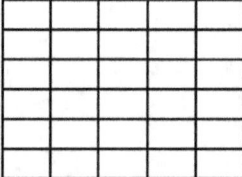

Pictorial Reasoning   Answers-125
© Gift Of Logic, Inc * Copying prohibited

Name _____   Date _____

SPOT THE DIFFERENCE

Spot the difference between the two figures in the questions below.

9

10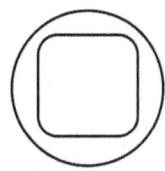

11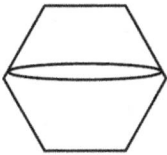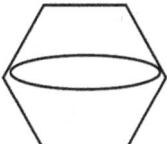

12

Pictorial Reasoning   Answers-125
© Gift Of Logic, Inc * Copying prohibited

ANSWERS

FINDING THE TRUTH

Q#	Answer	Reasoning
1	B-False	Not all flowers are red. Some are blue, purple etc.
2	B-False	We breathe in oxygen and breathe out carbon dioxide.
3	A-True	Sun is at the center of our Solar system.
4	B-False	Normal body temperature is 98.6 degrees Fahrenheit, not 89.6 degrees Fahrenheit.
5	A-True	Pediatricians specialize in taking care of children.
6	B-False	Earth has two thirds water and one third land.
7	A-True	Alexander Graham Bell is the inventor of telephone.
8	B-False	Plumbers are trained to fix problems with pipes. Electricians fix problems with electricity.
9	B-False	When it rains, the roads are wet.
10	A-True	Check the world map to verify this fact.
11	A-True	Ostrich is a bird that cannot fly. Research to verify this fact.
12	B-False	It is a proven fact that Sun is at the center of the solar system.
13	A-True	Paris is the capital of France, not U.S.A.
14	A-True	Fish can not survive on land.
15		Light _is_ needed in order to see. To sing a song, _it is not_ necessary to be a famous singer.

Answers

© Gift Of Logic, Inc * Copying prohibited

WORD ANALOGY

Q#	Answer	Reasoning
1	B-plants	Zoology is the study of animals. Botany is the study of plants.
2	A-drive	You ride a motorbike. You drive a car.
3	B-play	You read a book. You play a guitar.
4	B-drought	Too much rain causes flood. Too much Sun causes drought.
5	B-poor	Affluent means being rich. Impoverished means being poor.
6	B-inside	Foreign is something that is outside. Native refers to something that is inside.
7	A-North America	Singapore is in Asia. Canada is in North America.
8	A-Art	English is a language. Painting is an art.
9	A-movie	You get food in a restaurant. You see movies in a theater.
10	A-breathe	A stomach's job is to digest. The job of lungs is to breathe.
11	B-hear	Light is used to see. Sound is used to hear.
12	B-feet	Roof and foundation are at the top and bottom of a building. Similarly, the head is at the top and the feet is at the bottom.

Answers

© Gift Of Logic, Inc * Copying prohibited

SYNONYMS/ANTONYMS

Q#	Answer	Reasoning
1	B-False	To mute the sound is to turn off the sound.
2	B-False	To be enthusiastic is to be eager, not bored.
3	B-False	A quack is a person who cheats. Cheats are not respected.
4	A-True	A gullible person is easily cheated. A naive person is also easily cheated.
5	B-False	To reject something is the same as discarding it.
6	B-False	To dodge is to avoid. Dodging responsibility is the same as avoiding responsibility.
7	A-True	Trespass means to enter someone's property without permission. This is the same as invasion.
8	B-False	Fluctuate is the opposite of steady.
9	B-False	To retrieve something is to get it, not lose it.
10	A-True	To be feeble is to be weak, which is not the same as being strong.
11	A-True	Fulfilling and finishing are synonyms.
12	B-False	Shoving is the same as pushing - not pulling.
13	B-False	Snapping is the same as cutting. This is not the same as stretching.

Answers

AGREE-DISAGREE

1

Dave: The big circle is... Diane: The small circle is...
Answer: A) agree with each other.

Reasoning: Obviously both mean the same thing, but explain it in different ways. Dave says the big circle is outside the small circle and Diane says that the small circle is outside the big circle. Effectively, both say the same thing.

2

Fred: This gold chain...   Harry: This gold chain...
Answer: A) have the same opinion about the gold chain.

Reasoning: Harry thinks that the gold chain is not cheap which is the same as thinking that it is expensive. Fred also thinks it is expensive. So, they are both in agreement.

3

Harry: This book is full.. Larry: I was not tense..
Answer: B) disagree about the amount of suspense in the book.

Reasoning: Harry says the book was full of suspense, while Larry says that the book had very little suspense. So, they are clearly in disagreement.

Answers

© Gift Of Logic, Inc * Copying prohibited

AGREE-DISAGREE

4

Mark: Spending too much... Mary: Saving very little...
Answer: A) agree with each other.

Reasoning: Mark is of the opinion that spending too much money is not a good habit. If one spends too much money, then they are going to save very little. So, Mark is essentially saying that saving very little money is not a good habit. Mary also says the same thing.

5

Rohan: This car is not fit.. Randy: This car has safety..
Answer: A) agree with each other.

Reasoning: Randy says that the car has safety problems. It is common sense that a car with a safety problem is not fit for driving. This is the opinion that Rohan has. So, they are in agreement with each other.

6

Paul: Tall people live longer.. Kevin: Being tall or short..
Answer: B) disagree with each other.

Reasoning: Paul is of the opinion that tall people live longer than short people. Kevin is of the opinion that it does not matter whether one is tall or short when it comes to how long they will live. These are different opinions.

Answers
© Gift Of Logic, Inc * Copying prohibited

AGREE-DISAGREE

7

Anita: Some insects bite.   Andrea: All insects bite.
Answer: B) disagree with each other.

Reasoning: While Andrea says all insects bite, Anita says that some insects bite - thereby indicating that not all insects bite. They are not in agreement.

8

Brad: Many cities... Jennifer: Only a few cities..
Answer: B) disagree with each other.

Reasoning: Brad says "many cities" while Jennifer says "only a few cities". Clearly, these are opposites and therefore, they are in disagreement.

9

Neil: We should always explore.. Lance: We should never stop..
Answer: A) agree with each other.

Reasoning: Neil says that we should "always" explore the universe. Lance says that we should "never stop" exploring the universe - that is, we should "always" explore the universe. So, both of them agree.

Answers
© Gift Of Logic, Inc * Copying prohibited

AGREE-DISAGREE

10

Cathy: At least five fire trucks... Calvin: At most five trucks...
Answer: B) disagree with each other.

Reasoning: When Cathy says that at least five trucks are needed, she means that it is possible that more than five trucks could eventually be required. When Calvin says that at most five trucks are needed, he means that there will not be a need for more than five trucks. So, they are in disagreement over the number of trucks needed.

11

Liz: All the students.. Eric: None of the students..
Answer: A) agree with each other.

Reasoning: None of the students being absent is the same as all the students being in attendance. So, there is agreement between the two.

12

Fred: I go to restaurants... Farida: You rarely...
Answer: B) disagree with each other.

Reasoning: Rarely going to restaurants is not the same as frequently going to restaurants. So, there is disagreement between the two.

Answers
© Gift Of Logic, Inc * Copying prohibited

INFERENCE

1 Viral infections ...

Answer: A <u>Reasoning:</u> The statements say that viral infections cannot be treated, but bacterial infections can be. Stacy also has a bacterial infection in her throat. Therefore, we can infer that Tracy's throat infection can be treated with antibiotics.

2 The policemen are looking for ..

Answer: B <u>Reasoning:</u> Careful reading will help to identify the color of the pants and shirt worn by Mr. Rogers. The policemen were looking for a man with a red shirt and blue pants. Since they arrested Mr. Rogers, we can infer that Mr. Rogers was wearing a red shirt and blue pants.

3 The closer a planet is ...Answer: A

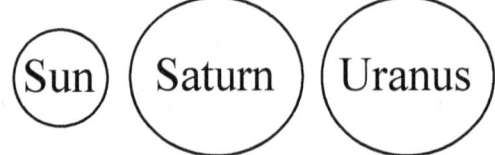

<u>Reasoning:</u> The first statement says that " the closer a planet is to the Sun, the warmer it is". The second statement says that Saturn is closer to Sun than Uranus. Therefore, we can infer that Saturn is warmer than Uranus. You can draw a picture to visualize the facts and make your inference.

4 Javed is in the same class ... Answer: B

(Javed Tina Randy)   <u>Reasoning:</u> Drawing a circle and writing the names will help with inferencing. Javed is in the same class as Tina. Tina is in the same class as Randy. So, we can infer that Javed is in the same class as Randy. So, choice B is the correct answer.

Answers

© Gift Of Logic, Inc * Copying prohibited

INFERENCE

5 For the Independence day .. Answer: A

Reasoning: Keep the facts in mind -Blue or Green pants, and Red or White shirts are allowed. So, we can infer from the facts that it is okay to wear a blue pant and a white shirt.

---

6 There are three lanes.. Answer: C

Reasoning: Note that one of the lanes is closed due to construction. Therefore, we can infer that two lanes are open. We cannot infer whether the left most lane is open or closed from the given statements.

---

7 Except Boris... Answer: B

Reasoning: Understanding the meaning of the word "except" is important to answer this question. Boris did not submit the homework and did not get a prize. Everyone except Boris got a prize.

---

8 In addition to Javed... Answer: B

Reasoning: The statement mentions that "in addition to Javed, three others..". So, we can infer that Javed is not the only one to get a gold medal.

Answers

© Gift Of Logic, Inc * Copying prohibited

INFERENCE

9 Magicians perform .. Answer: B.

<u>Reasoning</u>: Write the statements with arrows as shown to describe the facts clearly. Then, it is easy to make the correct inference.

  People who perform tricks → smart
  Magicians perform tricks.

So, since magicians are people, we can infer that magicians are smart.

---

10 Small boats... Answer: B

<u>Reasoning</u>:

  Boats with small fuel tank → cannot travel long distance
  Small boats have a small fuel tank.

So, small boats cannot travel long distance. This is the same as saying that small boats can travel short distances only.

11

Cats eat rats.. Answer: A

<u>Reasoning</u>: Follow the chain of arrows to make the inference. The arrow here means "eat".

  Cats → (eat) rats. Rats → (eat) bugs.
  So, cats → (eat) bugs.

12

We get sunlight .. Answer: A

<u>Reasoning</u>: Follow the chain of arrows to make the inference. The arrow here means "gives".

  Sun → Sunlight   Sunlight → Energy
  Therefore, Sun → Energy.

Answers

© Gift Of Logic, Inc * Copying prohibited

INFERENCE

13 Cake is made with eggs ... Answer: A

<u>Reasoning:</u> Note that the correct answer is the one that cannot be true.
Cake is made with eggs.
Cost of eggs is going to increase soon.
Therefore, the cost of cake (which has eggs) will also increase soon.
Therefore, it cannot be true that the cost of cake will decrease soon.

---

14 Driving in the wrong lane is illegal. Illegal driving is dangerous. Answer: A

<u>Reasoning:</u> Driving in wrong lane → illegal   illegal driving → dangerous
Therefore, by chaining, we can say that driving in the wrong lane is dangerous. Choice A says that driving in the wrong lane is not dangerous and hence it cannot be true. The answer choice that cannot be true is the correct answer.

---

15 Martin is four feet..   Answer: A

<u>Reasoning:</u>   Everyone taller than three feet can dive.
   Martin is more than three feet tall.
   So, Martin can also dive.
Choice A says that Martin cannot dive in the swimming pool. This cannot be true and therefore, this is the correct answer.

---

16 When water and ink are mixed .. Answer: A

<u>Reasoning:</u> Since water takes the color of the ink, when red ink is mixed with water, water will turn red. Choice A cannot be true because it says that water turned blue.

Answers
© Gift Of Logic, Inc * Copying prohibited

SORTING

1

Answer: B) False.

Reasoning: If the list is sorted in ascending order, then London must be the first in the list.

2

Answer: A) True.

Reasoning: Letters T, V, W and Z are ascending order of alphabetic sequence.

3

Answer: B) True.

Reasoning: C, E, K and W (first letters for Chaplin, Ewing, Kazi and Wadia) are in ascending order.

4

Answer: B) False.

Reasoning: S, T, A and N (the first letters of Steve, Tom, Abdul and Nilesh) are not in descending order.

Answers

© Gift Of Logic, Inc * Copying prohibited

ASSORTED PROBLEMS

1

Cities named A, B, C, D ...

Drawing a picture of the connections between the cities will help in answering the questions correctly.

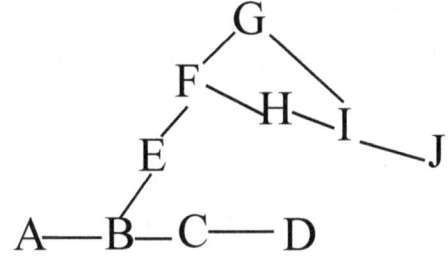

1) To go from A to H, we must go through G.
Answer: B) False. It is not necessary to go through G. We can go from A to F and then from F to H.

2) To go from A to H, we must go through F.
Answer: A) True. There is no other way to go from A to H.

3) To go from A to I, we must go through H.
Answer: B) False. We can go from A to I via A-B-E-F-G-I and avoid H.

4) The number of ways in which we can go from A to J is
Answer: B) 2. One way to go from A to J is A-B-E-F-H-I-J and another way is A-B-E-F-G-I-J.

Answers
© Gift Of Logic, Inc * Copying prohibited

ASSORTED PROBLEMS

2

1) How many green buses should you take to go from start to end?
Answer: B) 4. Reasoning: There are four lanes that go east or west. (horizontal lanes)

2) How many red buses should you take to go from start to end?
Answer: B) 5. Reasoning: There are five lanes that go north or south. (vertical lanes)

3) How many buses should you take to go from start to end?
Answer: C) 9. Reasoning: 4 green buses and 5 red buses add up to 9.

3

| parrot |
| eagle |
| dove |
| peacock |
| flamingo |

| tiger |
| lion |
| fox |

Answer: B) 5 birds as shown above.

---

4   A book shop carried ..

1) Jane took 9 old books.. How many new books did she get?

Answer: B) 6. Reasoning: For every 3 old books, you get 2 new books. So, for the 9 old books that Jane took, she will get 6 new books.

Answers
© Gift Of Logic, Inc * Copying prohibited

ASSORTED PROBLEMS

4 2) Jack took 5 old books.. How many new books did he get?
Answer: A) 2. <u>Reasoning:</u> For three old books, Jack got 2 new books in exchange. The other two books that he took with him could not be exchanged.

5

Foreign exchange rates..
1) Meghan went to India.. What did the bank swap with her?
Answer: B) 400 Rupees. <u>Reasoning:</u> 1 U.S dollar can be exchanged for 40 Indian Rupees. So, 10 U.S dollars can be exchanged for 400 Rupees.

2) Mathew went to London..
Answer: A) 200 Pounds. <u>Reasoning:</u> 1 U.S dollar can be exchanged for 2 British Pounds. 100 U.S dollars for 200 Pounds.

6

Swap the wires in the left column so that their colors match their slots in the right column.

red wire	yellow wire - swap 1	yellow slot
blue wire	green wire - swap 2	green slot
green wire	blue wire - swap 2	blue slot
yellow wire	red wire - swap 1	red slot

<u>Reasoning:</u> The yellow wire and the red wire are swapped to match the yellow slot and the red slot. Then, the blue wire and the green wire are swapped to match the blue slot and the green slot.

Answers
© Gift Of Logic, Inc * Copying prohibited

ASSORTED PROBLEMS

6

1) How many swaps does it take.. Answer: B) 2.

Reasoning: In the first swap, the red wire and the yellow wire can swap their positions, which will put the red wire in the red slot and the yellow wire in the yellow slot. In the second swap, the blue wire and the green wire can be swapped. After these two swaps, all the wires will be in correct slots that match their colors.

7

position	object	swap
1	apple	banana
2	orange	orange
3	banana	apple

8

Swap Amber's position with Brendan's. Swap Drew's position with Kate's. Write the names in the new positions.

Brendan	Kate
Drew	Amber

9 Note that the row with numbers only is the one that has 8,7 and 9. The second row has both numbers and alphabets and is not swapped.

a	f	m
t	4	s
8	7	9

Answers
© Gift Of Logic, Inc * Copying prohibited

ASSORTED PROBLEMS

10 A book was bound ...

Answer: B) to reverse the order of the pages and bind again.
<u>Reasoning:</u> If the order of the pages are reversed, then page 1 will be in the beginning, followed by page 2 and so on.

---

11 Andrew is driving...
Answer: B) Reverse at Maple and Central. He cannot reverse at Central and Elm because Elm is one-way going the other way from the landmark. He must take a left at Maple and reverse direction at Maple and Central.

---

12 One answer is shown below. Is your answer the same?

2	construct walls and roof
1	lay foundation
3	install windows and doors
4	install carpets
5	paint outside

---

13 Chris plans to visit..

Answer: B) CB, BA, AD, DB, BC
<u>Reasoning:</u> Since the bag must come from aunt Anne to aunt Drew, he must visit aunt Anne before aunt Drew. Since aunt Drew wants to add something to the bag to give to aunt Baker, Christ must visit aunt Drew before he visits aunt Baker. So, he must visit aunts Anne, Drew, and Baker in that order. In other words, he must visit cities A, D and B in that order. Only choice B satisfies this condition.

Answers

© **Gift Of Logic, Inc * Copying prohibited**

ASSORTED PROBLEMS

14 Zoya's schedule for the evening is to be decided from the events shown:

    TV   5 PM - 5:30 PM
    Soccer   6 PM - 7 PM
    Movie   6:30 PM - 8:30 PM

1) Answer: B) Movie.
Reasoning: If she does not want to skip soccer, then she must play from 6 PM to 7 PM. This means that she must skip the movie which begins at 6:30 PM.

2) Answer: B) Soccer.
Reasoning: If she wants to watch a movie from 6:30 PM to 8:30 PM, she will have to skip soccer that begins at 6 PM and ends at 7 PM.

15
Victor's schedule for...
1) His uncle must postpone his visit by   Answer: B) 1 hour.
Reasoning: Victor is busy till 11 AM. So, his uncle has to postpone his visit by at least one hour - so, at 11:20 AM, Victor will be free to meet his uncle.

2) Answer: A) 7:15 AM.
Reasoning: Victor's math class beings at 8 AM. So, he is not free to meet his friend at 8 AM. But, he is free before 8 AM. So, his friend must advance his visit to 7:15 AM so that after 45 minutes, Victor can attend his math class at 8 AM.

Answers
© Gift Of Logic, Inc * Copying prohibited

| POSITIONING | FIXED |

1

Tom, Dick, and Harry must be seated in three stools. One in each stool. Tom must sit in stool #1. Write this rule as Tom-1 so that you can remember and apply this rule when you answer the questions.

Which of the following seatings are correct?
As you look at each of the choices, look for Tom-1(that is, look for Tom in stool# 1).

A) Incorrect - because you cannot find Tom-1.

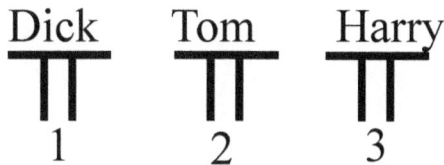

B) Correct - you can find Tom-1.

C) Correct - you can find Tom-1.

Tom   Harry   Dick
 ⊓   ⊓   ⊓
 1   2   3

Can Harry sit in stool# 1?   Answer: B) No. Only Tom can sit in stool# 1.

Answers

POSITIONING   OR

2   Andy, Brittany, and Cathy must be seated..

Andy can sit only in chair# 1 or chair# 3 . Write this rule as Andy-1 or Andy-3 and remember it as you answer the questions.

A) Incorrect - cannot find Andy-1 or Andy-3.

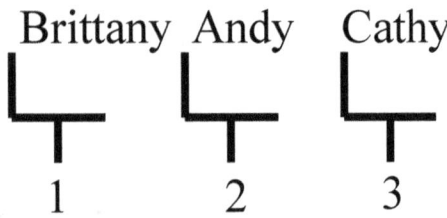

B) Correct - can find Andy-1.

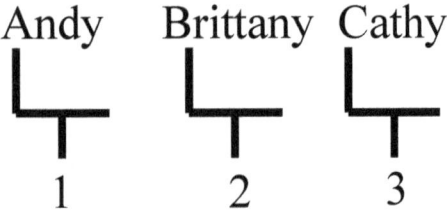

C) Correct - can find Andy-3.

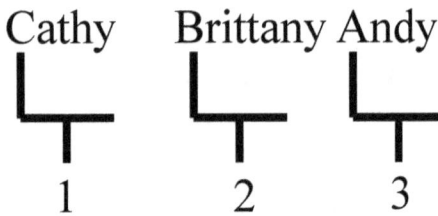

D) Incorrect - Andy cannot sit in two chairs.

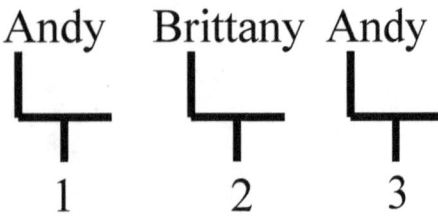

Andy can sit in any of the three chairs. Answer: B) False. Andy cannot sit in chair# 2.

Answers
© Gift Of Logic, Inc * Copying prohibited

POSITIONING — LEFT OF

3

Andy, Brittany, and Cathy must be seated..

Andy must sit to the left of Brittany. Write this rule as Andy-Brittany and look for it as you answer the questions. The dash '-' means that Andy and Brittany can sit next to each other, or someone can sit between them.

A) Incorrect - cannot find "Andy-Brittany". Andy is sitting to the right of Brittany.

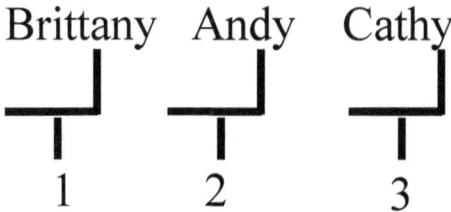

B) Correct- can find "Andy-Brittany".

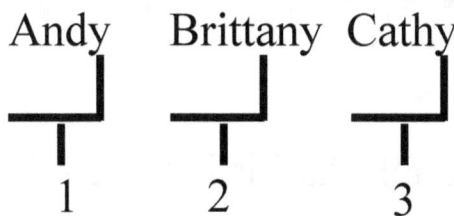

C) Correct - can find "Andy-Brittany", even though Cathy is sitting in between them.

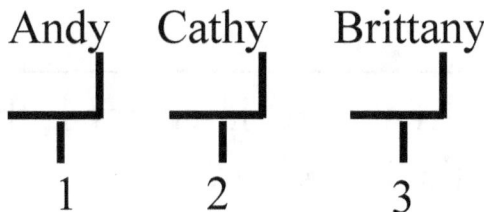

Answers

| POSITIONING | RIGHT OF |

4

Paul, Queenie, and Robert must be seated in the three chairs; only one in each chair. Queenie must sit to the right of Robert.

Write this rule as Robert-Queenie and look for this seating arrangement in the choices.

A) Incorrect - cannot find Robert-Queenie.

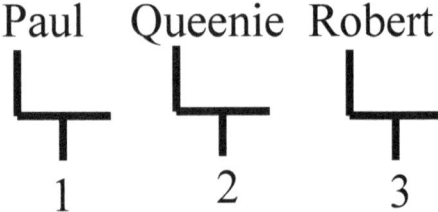

B) Correct - Can find Robert-Queenie.

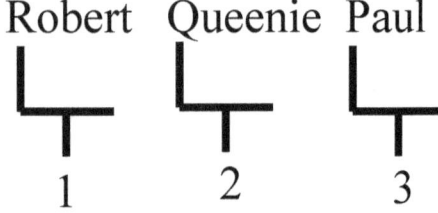

C) Correct - can find Robert-Queenie, even though Paul is between them.

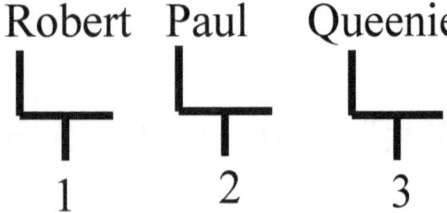

Can Queenie sit in chair #1? Answer: B) No. If she sits in chair #1, then Robert cannot sit to her left, since chair #1 is the first chair.

Answers

| POSITIONING | BETWEEN |

5 Larry, Margie, and Nina are to be seated in three chairs.

Larry must sit between Margie and Nina. Write this rule as "Larry between " and look for this seating arrangement in the choices. Note that Margie and Nina can be on either side of Larry.

Which of the following seatings are correct?

A) Incorrect - cannot find "Larry between" seating arrangement.

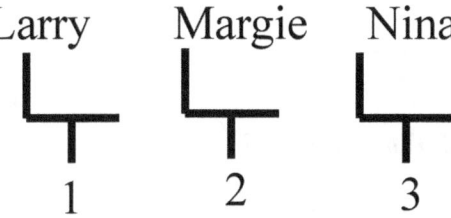

B) Correct - can find the "Larry between " seating arrangement.

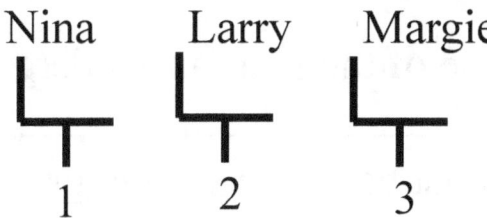

C) Correct - can find the "Larry between" seating arrangement.

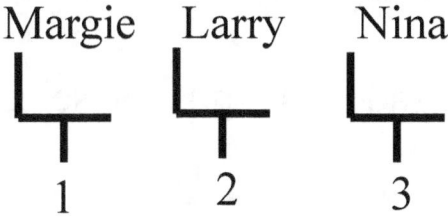

Larry can sit only in chair # 2. Answer: A) True. This is the only way he can sit between Margie and Nina.

Answers
© Gift Of Logic, Inc * Copying prohibited

PICKING

The group below shows five students and their grades.

1

Larry	Kindergarten
Mary	First grade
Nancy	Second grade
Oliver	First grade
Patrick	Kindergarten

From the above group of students, a team of three students is to be picked. Two of them must be in kindergarten.

Which of the following choices represents a correct team?

A) Larry, Mary, Oliver >> Incorrect - Only one of them is in kindergarten.
B) Larry, Patrick, Oliver >> Correct team.
C) Mary, Nancy, Oliver >> Incorrect - None of them are in kindergarten.

Pick a team of three students. One student must be in the second grade.

Which of the following choices represents a correct team?

Since Nancy is the only student in the second grade, she must be selected in the team.

A) Nancy, Larry, Mary >> Correct .
B) Mary, Oliver, Patrick >> Incorrect - Nancy is not in the team.

Answers

PICKING

2

Five balls are available in the colors shown above.
A set of three balls must be picked from the five.
Each ball must be of different color.
Which of the following choices represent a correct set?

A) (Red) (Green) (Blue) >>Correct: Each ball is of different color.

B) (Red) (Blue) (Green) >>Correct: Each ball is of different color.
                          The order of picking does not matter.

Five balls are available in the colors shown above.
A set of three balls must be picked. Two balls must be of the same color.
Which of the following choices represent a correct set?

A) (Red) (Red) (Blue) >>Correct: There are two red balls.

B) (Red) (Green) (Blue) >>Incorrect: All the balls are different.

Answers   101
© Gift Of Logic, Inc * Copying prohibited

PICKING

3

The group below shows four students,
Andy, Brittany, Cathy, Derek

Pick a team of three students. Andy and Brittany must be picked together.

Which of the following choices represents a correct team?

A) Andy, Brittany, Cathy >> Correct - Andy and Brittany are together.

B) Andy, Cathy, Derek >> Incorrect - Andy and Brittany are not together.

C) Derek, Brittany >> Incorrect - There should be three students in the team.

D) Cathy, Andy >> Incorrect - There should be three students in the team.

E) Brittany, Andy, Derek >> Correct - Andy and Brittany are together.

Answers

PICKING

4

P, Q, R, S, and T are the names of five students.
Two teams, team A and team B are to be formed out of these five students.

Write the rules in short form to help you remember them.

Team A must have three students. >> A3
Team B must have two students. >> B2
P and S must be in Team A. >> P,S - A
Q and R must be in team B. >> Q,R - B

Which of the following choices represent the correct teams?

Apply the rules to each choice.

<u>Team A</u>   <u>Team B</u>

A) P, S, T   Q, R   >>Correct, meets all the rules.

B) P, S   Q, R, T >>Incorrect - Team A must have three students.
                         A3 rule is not met.

C) Q, R, T   P, S   >>Incorrect - P,S-A rule not met.
                         P and S are not in team A.

Answers
© Gift Of Logic, Inc * Copying prohibited

PICKING

5

L, M, N, O, and P are the names of five students.
Two teams, team A and team B are to be formed out of these five students.

Write the rules that are given in short form to help you remember them.

Team B must have two students. >> B2

M and N must be in the same team. >> M,N

L and O cannot be in the same team. >> L,O not in the same team.

Which of the following choices represent the correct teams?
Apply the rules to each choice.

	Team A	Team B
A)	M, N, O	L, P >>Correct teams.
B)	M, P, L	N, O >>Incorrect - does not meet M,N rule. N is missing from Team A.
C)	L, N, O	M, P >> Incorrect - L and O cannot be together.
D)	M, N	P >>Incorrect - does not meet the B2 rule. - two students must be in team B.

Answers

NUMERIC SUDOKU

1

1		4	3
4	3		1
2	1		4
3		1	2

1	2	4	3
4	3	2	1
2	1	3	4
3	4	1	2

2

4	3	1	
2		3	
1		2	3
	2		1

4	3	1	2
2	1	3	4
1	4	2	3
3	2	4	1

Answers

© Gift Of Logic, Inc * Copying prohibited

NUMERIC SUDOKU

3

	3	4	2
	4	1	3
3		2	
4	2	3	

1	3	4	2
2	4	1	3
3	1	2	4
4	2	3	1

4

2	1	4	3
4		1	2
1		3	
3	4	2	

2	1	4	3
4	3	1	2
1	2	3	4
3	4	2	1

Answers

NUMERIC SUDOKU

5

2	1	4	3
	3		
3	2	1	4
	4	3	

2	1	4	3
4	3	2	1
3	2	1	4
1	4	3	2

6

3			4
1	4		2
4			1
2		4	3

3	2	1	4
1	4	3	2
4	3	2	1
2	1	4	3

Answers 107
© Gift Of Logic, Inc * Copying prohibited

NUMERIC SUDOKU

7

4	3	1	2
2			
3	2	4	1
1			3

4	3	1	2
2	1	3	4
3	2	4	1
1	4	2	3

8

2		4	1
1	4		3
	1		2
3	2		4

2	3	4	1
1	4	2	3
4	1	3	2
3	2	1	4

Answers
© Gift Of Logic, Inc * Copying prohibited

NUMERIC SUDOKU

9

			3
1	3	2	
		3	
3			1

2	4	1	3
1	3	2	4
4	1	3	2
3	2	4	1

10

			4
3		1	
	1	2	
2		4	

1	2	3	4
3	4	1	2
4	1	2	3
2	3	4	1

Answers

NUMERIC SUDOKU

11

3		4	
1	4		
	1	2	3
	3	1	

3	2	4	1
1	4	3	2
4	1	2	3
2	3	1	4

12

	4		
	2	3	4
2	3	4	
		2	

3	4	1	2
1	2	3	4
2	3	4	1
4	1	2	3

Answers

© Gift Of Logic, Inc * Copying prohibited

NUMERIC SUDOKU

13

	2		
1			4
		1	
3			2

4	2	3	1
1	3	2	4
2	4	1	3
3	1	4	2

14

3	4		
2			3
1			4
		1	

3	4	2	1
2	1	4	3
1	2	3	4
4	3	1	2

Answers
© Gift Of Logic, Inc * Copying prohibited

NUMERIC SUDOKU

15

	3		
	4	2	3
4	2	3	
		4	

2	3	1	4
1	4	2	3
4	2	3	1
3	1	4	2

16

3			2
	4	1	
	2	3	
1			4

3	1	4	2
2	4	1	3
4	2	3	1
1	3	2	4

NUMERIC SUDOKU

17

	2		
	3	1	2
2	4	3	
		2	

1	2	4	3
4	3	1	2
2	4	3	1
3	1	2	4

18

4	2		3
			2
2			
1	3		4

4	2	1	3
3	1	4	2
2	4	3	1
1	3	2	4

Answers
© Gift Of Logic, Inc * Copying prohibited

ALPHABETIC SUDOKU

1

	B		
			A
B		C	
	D		

A	B	D	C
D	C	B	A
B	A	C	D
C	D	A	B

2

			B
B		C	
	D		
C			A

D	C	A	B
B	A	C	D
A	D	B	C
C	B	D	A

Answers

© Gift Of Logic, Inc * Copying prohibited

ALPHABETIC SUDOKU

3

A			B
	A		
D		C	

A	C	D	B
B	D	A	C
C	A	B	D
D	B	C	A

4

B			
		A	
	B		D
C			

B	A	D	C
D	C	A	B
A	B	C	D
C	D	B	A

Answers
© Gift Of Logic, Inc * Copying prohibited

ALPHABETIC SUDOKU

5

B			C
	C		
	D	C	

B	A	D	C
D	C	B	A
C	B	A	D
A	D	C	B

6

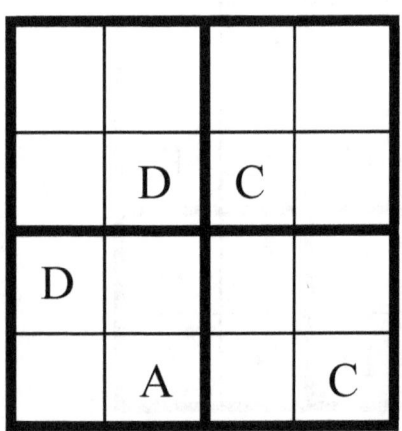

C	B	A	D
A	D	C	B
D	C	B	A
B	A	D	C

Answers 116
© Gift Of Logic, Inc * Copying prohibited

ALPHABETIC SUDOKU

7

D			
		C	
			A
A	D		

D	C	A	B
B	A	C	D
C	B	D	A
A	D	B	C

8

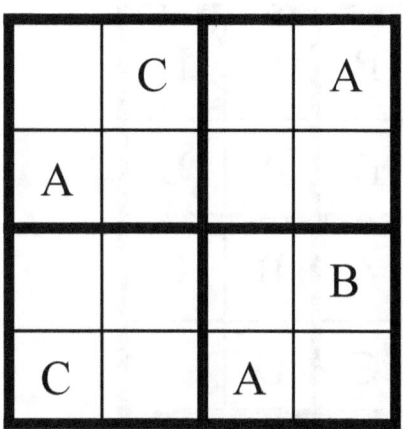

B	C	D	A
A	D	B	C
D	A	C	B
C	B	A	D

Answers
© Gift Of Logic, Inc * Copying prohibited

ALPHABETIC SUDOKU

9

			C
A		B	
		C	
C			A

B	D	A	C
A	C	B	D
D	A	C	B
C	B	D	A

10

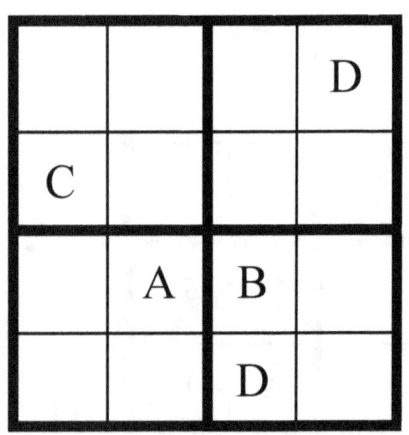

A	B	C	D
C	D	A	B
D	A	B	C
B	C	D	A

Answers 118
© Gift Of Logic, Inc * Copying prohibited

ALPHABETIC SUDOKU

11

C		D	
A		C	
	A		C
	C		D

C	B	D	A
A	D	C	B
D	A	B	C
B	C	A	D

12

	D		
	B	C	D
B	C	D	
		B	

C	D	A	B
A	B	C	D
B	C	D	A
D	A	B	C

Answers

© Gift Of Logic, Inc * Copying prohibited

ALPHABETIC SUDOKU

13

	B		
			D
		A	
C			B

D	B	C	A
A	C	B	D
B	D	A	C
C	A	D	B

14

	D		
			C
A			
		A	

C	D	B	A
B	A	D	C
A	B	C	D
D	C	A	B

Answers 120

© Gift Of Logic, Inc * Copying prohibited

ALPHABETIC SUDOKU

15

	C		
	D	B	C
D	B	C	
		D	

B	C	A	D
A	D	B	C
D	B	C	A
C	A	D	B

16

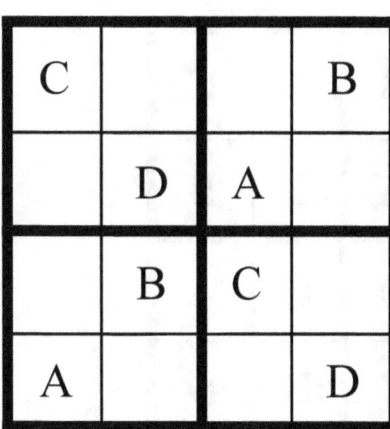

C	A	D	B
B	D	A	C
D	B	C	A
A	C	B	D

Answers
© Gift Of Logic, Inc * Copying prohibited

ALPHABETIC SUDOKU

17

	B		
	C	A	B
B	D	C	
		B	

A	B	D	C
D	C	A	B
B	D	C	A
C	A	B	D

18

D	B		C
			B
B			
A	C		D

D	B	A	C
C	A	D	B
B	D	C	A
A	C	B	D

Answers 122
© Gift Of Logic, Inc * Copying prohibited

PICTURE SEQUENCE

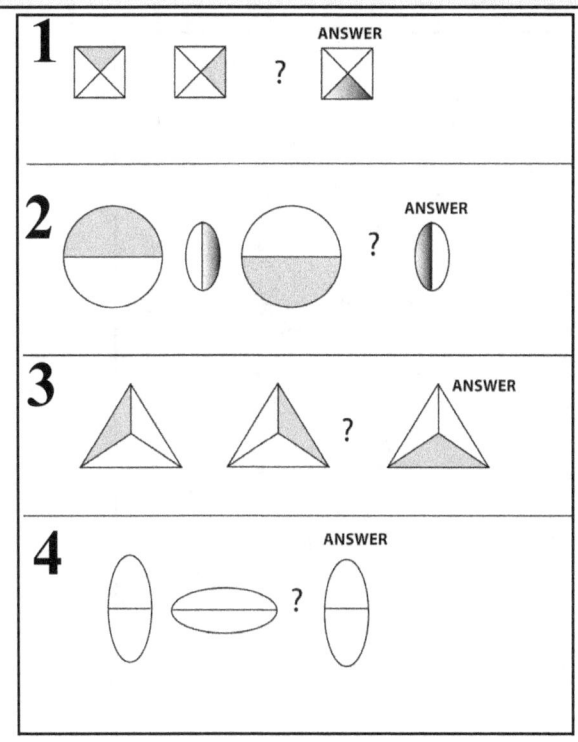

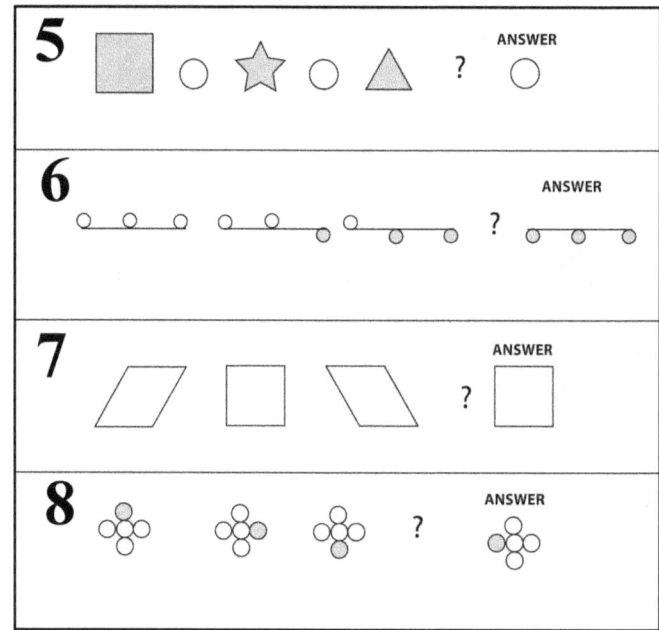

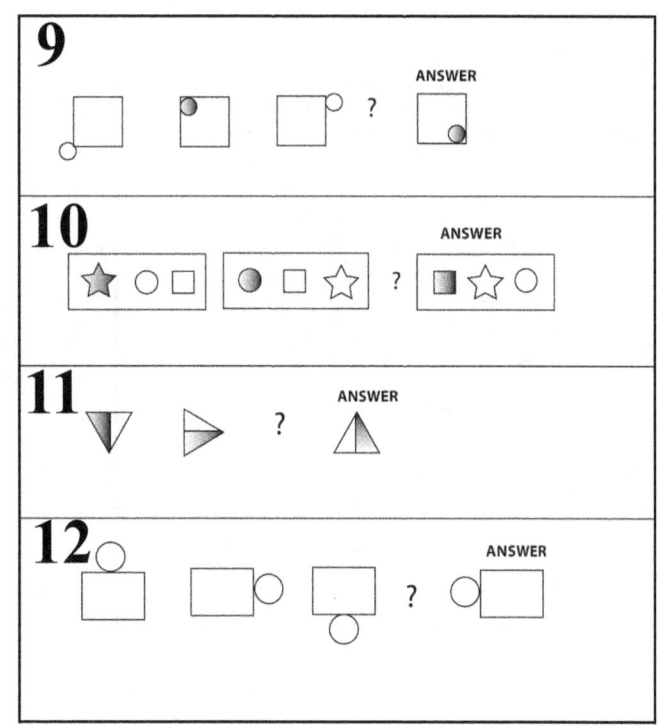

Answers 123
© Gift Of Logic, Inc * Copying prohibited

ODD PICTURE

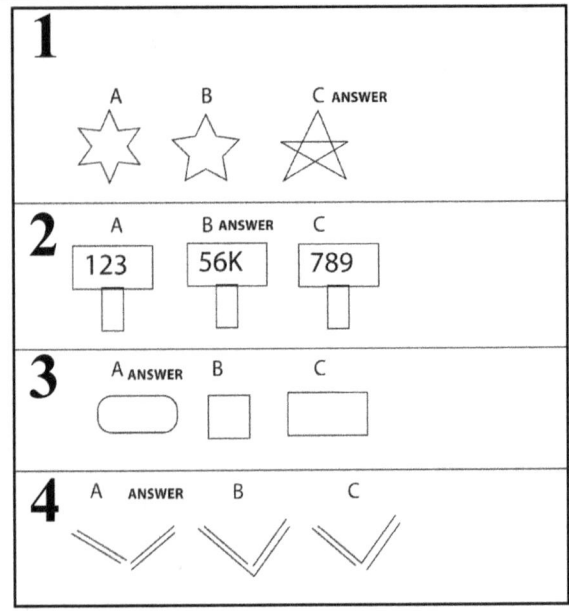
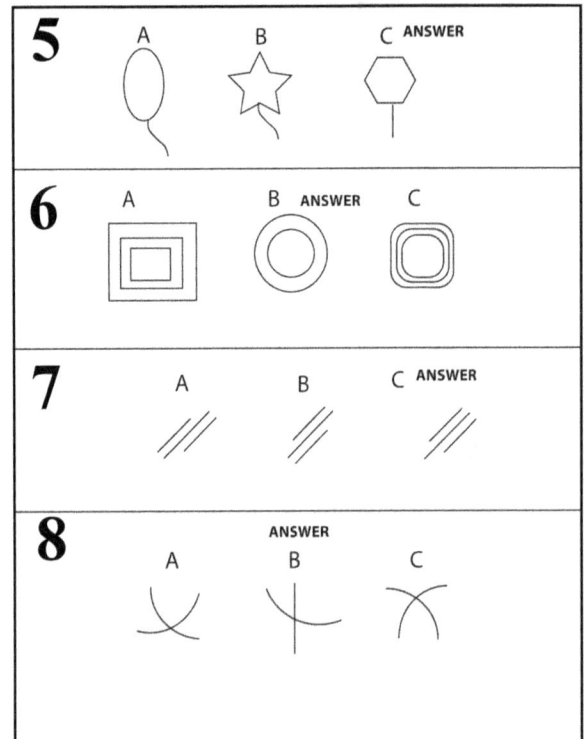
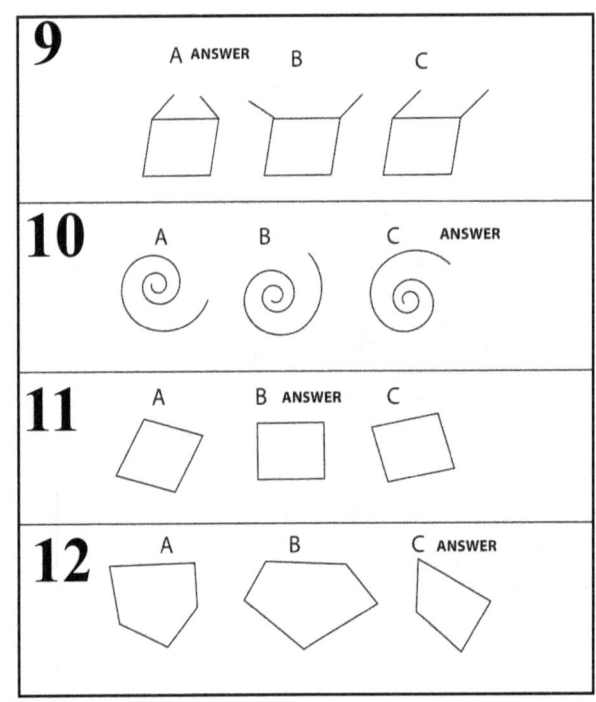

Answers
© Gift Of Logic, Inc * Copying prohibited

SPOT THE DIFFERENCE

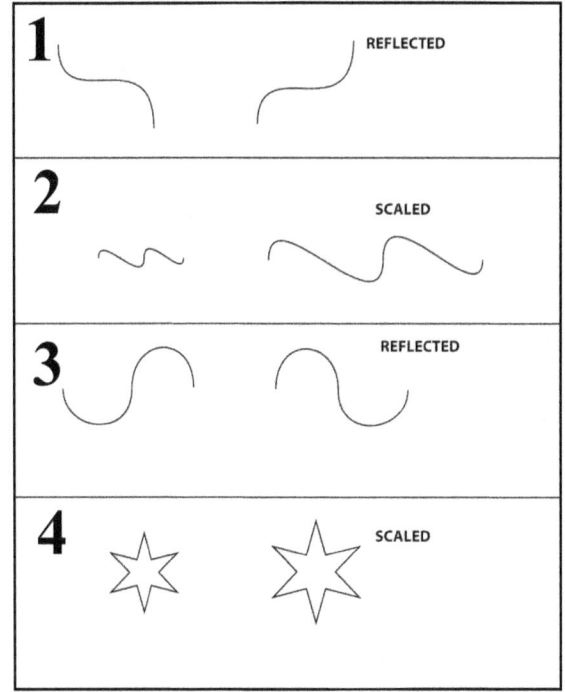

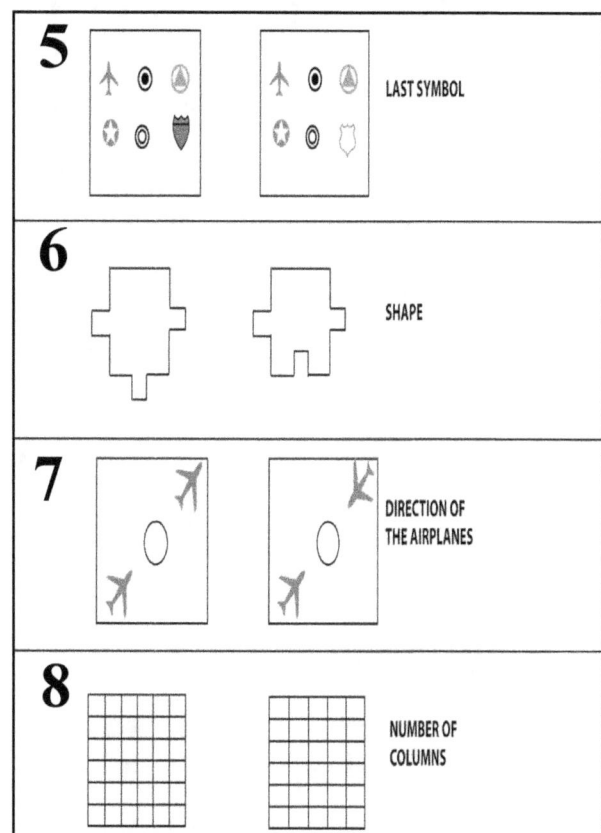

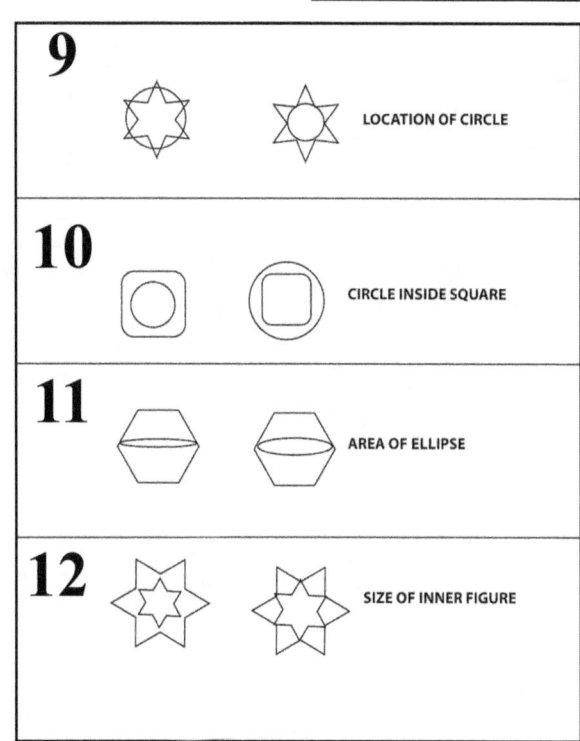

Answers 125
© Gift Of Logic, Inc * Copying prohibited

NOTES

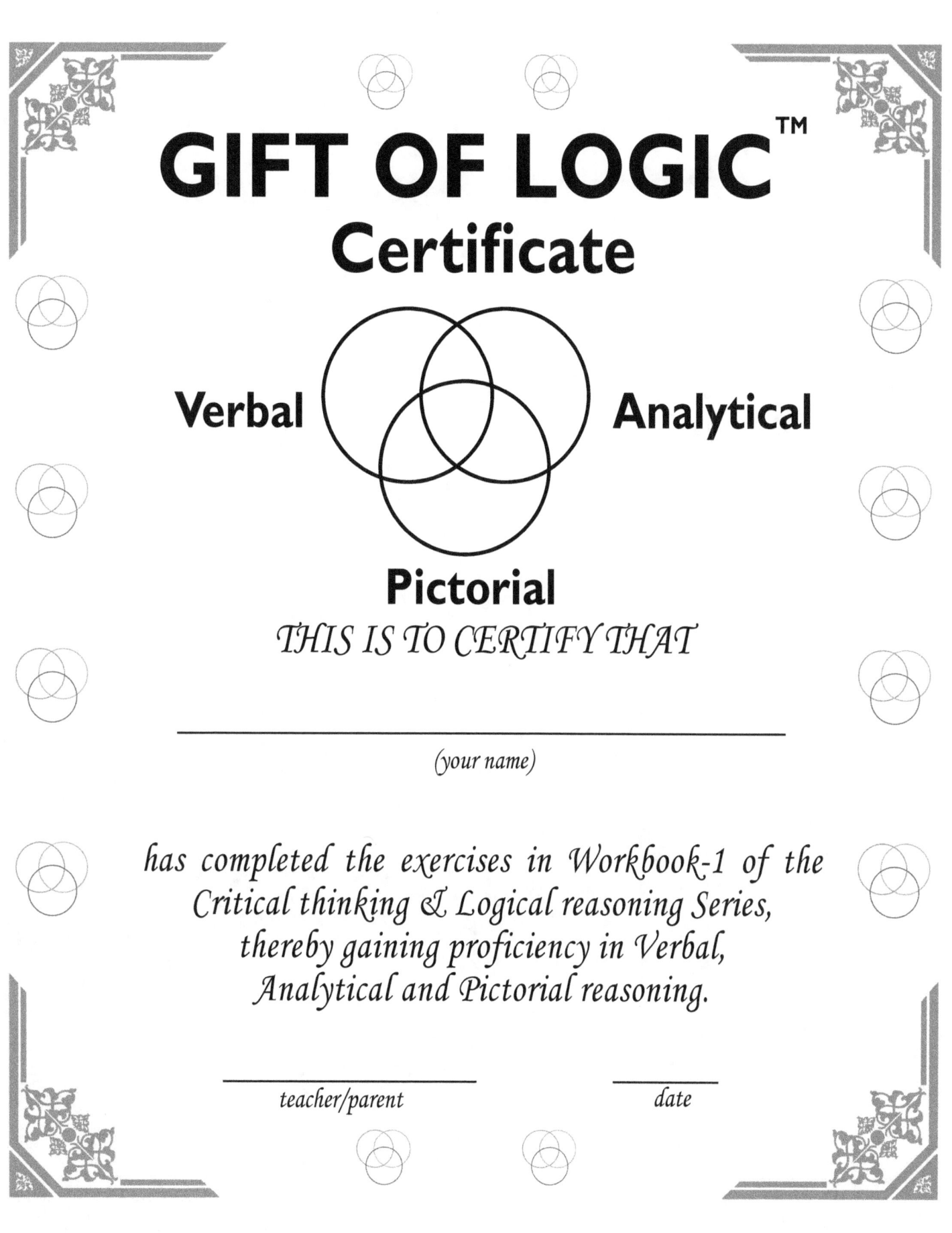

NOTES